"For we do not wrestle against flesh and blood, but against the rulers, against the authorities, against the cosmic powers over this present darkness, against the spiritual forces of evil in the heavenly places."

Ephesians 6:12

The Ephesians Hypothesis

A Biblical View of the Paranormal and Supernatural

John Daniel McWilliams

The Ephesians Hypothesis:
A Biblical View of the Paranormal and Supernatural

Published by Rose Rock Media.
Visit **www.roserockmedia.net** for more information.

Bonnie McWilliams, Assistant Editor

ISBN- 978-1-7333997-2-2

Contents

1 A Tale of Two Realms 1

2 Whatever Happened to the Supernatural? 5
Roots of Doubt 8
How Many Miracles Do We Need? 10

3 Supernatural Intrusions 12
Our "Closed" Universe? 15
The Gap 17
All in the Numbers? 19

4 Paranormal Pickles 22

5 The Ephesians Hypothesis 26
Living with the Unseen 28

6 Dwellers in the Supernatural Realm 33
The Most High 34
What Can't God Do? 36
Angels and Others 41
A Divine Council? 42
The Rise and Fall of Lucifer 44
A Fallen but Formidable Army 47
Spirits in the Earthly Realm 48
How Do Satan and His Cohorts Affect Us? 53
Unpacking the Ephesians Hypothesis 56
Test the Spirits 58

7 Signs and Wonders 60
Are Modern Miracles Biblical? 61
Miracles and Catholics 65
The Genuine Article 71
Discerning a True Miracle 73

8 Ghosts and Other Things that Go Bump in the Night 74
What Does a Ghost Look Like? 76
Bible Ghosts 78
Our Dual Nature 79
Ghosts in the Machine 80
Are Ghostly Phenomena Consistent with the Ephesians Hypothesis? 82
Goblins, Gremlins, and Ghouls 86

9 THE OCCULT 89
Séances and Psychics 91
Spiritualism 92
Witches, Warlocks, and Wicca 93
Satanism 94
Halloween and Holy Days 96
Voodoo and Santeria 99

10 BIGFOOT AND FRIENDS 101
Cryptozoology at Work 103
Classified Cryptids 103
Are There Undiscovered Species? 110
A Closer Look at Bigfoot 111
A Closer Look at Sea Monsters 115
Is Nessie a Dinosaur? 116
Nearing Trail's End 119

11 BIGFOOT ON THE FRINGE 121
Sasquatch Hops the Fence 123
The Monsters on Maple Street 124
Shapeshifters and Skinwalkers 125
More Humanoid Haints 127

12 UFOS: THE LIGHTS' SIDE 131
A Very Brief History of UFOs 132
The Research 134
That 5% 135
Case Studies 138
Clash of the Real-World Worldviews 147

13 UFOS: THE INSIDE 149
The Theology of ETs 150
The Rime of the Ancient Aliens 153
The Science of ETs 154
A Grand Scheme? 157

14 UFOS: THE OTHER SIDE 160
The Interdimensional Hypothesis 162

15 UFOS: THE DARK SIDE 167
The Hill Abductions 169
The "Perfect Abduction" 171
A Riddle, Wrapped in a Mystery, Inside a Spaceship 175

The Dirty Little Secret of Abduction Cases 178
Are UFO and ET Phenomena Consistent with the Ephesians Hypothesis? 181

16 A Paranormal Potpourri 188
Bridgewater Triangle 190
Skinwalker Ranch 193
Dancing with the Apparitions 195
The Quintessential Ghost 201

17 Could It Be Magic? 209
Magic in the Bible 210
Serious Magic 211
Modern Magic 213
The Magic Christians 214
Effectual Prayer 215
If the Fates Allow 216
Blessings and Curses 217
Superstitions 219

18 Prophecy and Divination 221
Prophecy vs. Divination 222
Nostradamus 224
The Modern Mantis 226
We Three Kings of Orient Are What? 232
Prophets in Church 233
Deception in Prophecy 235
Dreams 236
The Biblical Standard 237

19 Parapsychology and Other Mind Games 240
Déjà Who? 241
Extrasensory Deception 242
Hypnosis 245
Telekinesis 247
If You Could Read My Mind 248
Near-Death Experiences 249
Artificial Intelligence 255

20 A New Age of Old Ideas 258
Decloaking the New Age 260
All is One 262
Spirit Guides 265
Kabbalah 266
Reincarnation and Karma 268

Yoga 269
Meditation 273
The New Age and UFO Religions 276

21 MEGA-MIRACLES 278
Creation 281
The Fall 283
The Flood 287
Babel 291
An Unseen Babel..294
The Narrative Sequence Fallacy...296

22 HOLY GROUND 299
Ley Lines...301
Stonehenge 302
Nature Religions..305
Bermuda Triangle...306
Vortexes 308
Crop Circles 310
Feng Shui 312
Spiritual Geography 313

23 AN UNDISCOVERED COUNTRY 316

24 DECEPTION'S DEAD END...329
The Ultimate Paranormal Deception......................................330
The Believer's Supernatural Hope 337

25 CHAPTER ONE? 339
Carry On...341

APP A - EXPERIENCING THE SUPERNATURAL GOD 345
GLOSSARY
INDEX

1
A Tale of Two Realms

"...deep in the human heart there are these desires—to experience the supernatural, to escape death, to know love that we can never lose, to not age but live long enough to realize our creative dreams, to fly, to communicate with nonhuman beings, to triumph over evil." Timothy Keller [1]

"What no eye has seen, nor ear heard, nor the heart of man imagined, what God has prepared for those who love him." 1 Corinthians 2:9

Fantasies, legends, fairy tales, myths… They often tell the human story better than our history books. Some claim that our draw toward adventure, our yearning for the happily-ever-after, and our craving for mystery are crutches we make for an aimless, nihilistic world. I believe differently.

Our Creator could have placed us as spirit beings hovering in the endless space surrounding him, absorbed in loving communion. But he didn't…

I have no doubt this would have worked quite well, but in a burst of divine creativity, the Most High amazed his angelic observers. God embedded us, his image-bearers, in a Story—yet not as a mere afterthought. We were *specifically* designed

[1] Timothy Keller, *Hidden Christmas,* (New York: Penguin Books, 2018), 25.

to fit into the narrative, originating in the Creator's mind and playing out in a physical world. We can move, interact, imagine, sense, wonder, search, decide, and *live the Story*. Much of Jesus's teaching was in a story form. He was certainly aware that humans understand truths housed within narratives more easily than we do strings of characters in a databank.[2]

Like our Designer (who even entered his own Story), we're plot-oriented beings, enjoying quests and creativity. We're attracted to things that are unique, remarkable, and beyond the normal. Things that are unseen, yet right alongside the Story seem to tug at our souls. This isn't some malicious remnant of our fallen nature, rather, it's a dynamic part of our eternal, creaturely nature.

But unfortunately, like the rest of our bent traits, our imaginations have also been corrupted. We demand miracles while God values faith. We replace the truly supernatural with paranormal counterfeits. We seek a divine experience when what we need is a divine relationship.

Our culture, and Christians in particular, drift toward two extremes when confronted with the supernatural. Some seem embarrassed by biblical miracles and go to great lengths to provide naturalistic explanations. Others are lured by tabloid paranormal claims, far-removed from reality. So why another book on the paranormal? Why should Christians focus on such fanciful, sideline topics when there are so many urgent issues facing the church today?

I'll admit up front that I'm intrigued by the fringe. I like *Star Trek*. I wish there really were Klingons somewhere out there. I've watched every season of *Ghost Whisperer*, bad theology and all. I think somewhere there's a highland castle

---

[2] Some define this as a *metanarrative*, a "grand story" that gives meaning to the workings of our universe. This view is consistent with Scripture but is not popular in today's postmodern philosophy (see glossary).

with my name over the door. I hope I stumble into Bigfoot someday.

Perhaps this quirkiness is the very thing that has driven me over the years to ponder things outside of traditional science, yet still with rational moorings. Yes, I am a scientist—at least I play one at work. And I contend that my science training has worked quite well to balance what would otherwise become fanatical obsessions. My research (both fun and serious) into paranormal phenomena has revealed three timely issues...

1. A significant number of Christians sidestep the biblical view of the supernatural. These believers either dismiss the supernatural completely or embrace the secular, pop-culture mix of paranormal notions. Both extremes are enticements to a new brand of false theology.

I believe the Church has the responsibility of equipping Christians with tools to think cogently, not just about theology, but about all areas of engagement. We are urged to always be "prepared to make a defense to anyone who asks you for a reason for the hope that is in you."[3] As a scientist, I would never propose that we refuse to explore the unusual. That would be a death knell for empirical science. Scripture describes a supernatural realm that is interfaced with our physical domain. We should have the freedom to examine the evidence, and whatever the phenomenon, right-thinking—biblical, logical, and informed—*will* lead to truth.

2. Our culture, and particularly our mission field, has a non-biblical view of the supernatural, and thus, reality in general. The Apostle Paul knew when he presented the gospel to the Athenians, he was addressing a group with a load of supernatural and philosophical baggage. Understanding and accommodating their worldview was crucial in effectively evangelizing them. In Ephesians, Paul also provided for us a

---

[3] I Peter 3:15

basic and crucial view of reality—two realms, one natural and one supernatural, both interacting with each other.

3. Beliefs have consequences. Erroneous and unexamined ideas about the supernatural have become a significant factor in our world's misunderstanding of important concepts such as faith, ethics, and morality. Wrong-thinking about the paranormal leads even some believers away from biblical truth and into deception, fear, and despair.

In the following chapters, we'll examine selected paranormal phenomena that are relevant to the biblical worldview. I don't claim that the topics included are equally worthy of consideration. A few are trivial or "fringe" but nevertheless, metastatic in our culture. As my criteria for selection I asked: 1) Is the idea engaging a significant portion of the culture? 2) Does the idea affect worldviews? 3) Does the biblical worldview address the idea?

The goal of this book isn't necessarily to legitimize or solve paranormal mysteries, although as a scientist, I make no apologies for presenting my own hypotheses when appropriate. My hope is that this discussion will inform and strengthen the biblical view of our Story, and ultimately magnify its Author.

2
Whatever Happened to the Supernatural?

"Traditional Christian teaching has for centuries kept the unseen world at arm's length. We believe in the Godhead because there's no point to Christianity without it. The rest of the unseen world is handled with a whisper or a chuckle."
Michael Heiser[4]

"The mind which asks for a non-miraculous Christianity is a mind in process of relapsing from Christianity into mere religion." C.S. Lewis[5]

In the early 1800s, Thomas Jefferson created his own new and improved (in his view) harmony of the four Gospels.[6] In one of the first cut/paste word processing operations (literally—cutting up a Bible and reorganizing it) Jefferson produced *The Life and Morals of Jesus of Nazareth*, commonly called "The Jefferson Bible." He purposely omitted all miracles and supernatural references, concentrating on the moral teachings of Jesus. This has caused criticism over the past two centuries, although Jefferson claims his objective was simply to use the

[4] Michael Heiser, *The Unseen Realm: Recovering the Supernatural Worldview of the Bible,* (Bellingham, WA: Lexham Press, 2015).
[5] C.S. Lewis, "Miracles," *The Complete C.S. Lewis Signature Classics*, (New York: Harper Collins, 2002), 420.
[6] *Thomas Jefferson Encyclopedia*, www.monticello.org, 2019.

work in his personal devotions. Whatever the intent, it's clear that Jefferson was uncomfortable with the supernatural.

We would certainly expect the secular world to hold a wide range of beliefs about the supernatural.[7] There are plenty of disbelieving naturalists on one end of the spectrum with glassy-eyed mystics on the other and everything imaginable in-between. It's disturbing, however, when Christianity's response, like Jefferson's, is to slice out those "embarrassing" miracles and narrow the focus exclusively to the physical.

While talk shows and paperbacks make millions peddling the supernatural, mainline Christianity seems to treat it as a pimple on the face of the gospel. Some pastors (and even entire denominations) tend to teach "around" the miraculous... "Surely there's a natural explanation for the awkwardness of a talking donkey (Numbers 22) or a natural reason for the delay of a sunset (2 Kings 20). When the Red Sea parted (Exodus 14) wasn't that just an exaggeration of the annual drought cycle? And that big fish story (Jonah 1)—isn't it just an allegory?"

Christian discourse today seems quite pleased to limit biblical supernatural elements to a nebulous deity, a naturalistic creation, and an allegorical resurrection. I believe this fragmented view of the supernatural has seriously disabled the view of reality held by many Christians. Our hymns and creeds readily affirm a transcendent, intervening Creator, but on Monday, a vague once-upon-a-time creation event is more tolerable. We ease our "miracle-anxiety" with a clockwork, cause-and-effect physical universe that is comfortably explainable. When cornered by the supernatural, we often rush to make up arbitrary rules about it, then struggle to make them fit the visible universe. Bland, deconstructed scriptural miracles

[7] The TV series *The X-Files* offers a smorgasbord of paranormal scenarios. I'm not advocating binge-watching it all, but it does seem to be almost a metaphor for our current supernatural confusion. Throughout the hodge-podge collection of paranormal episodes, the viewer senses a common thread. Some believe, some don't, some "want to believe."

fit more smoothly into secular conversations, as we keep the supernatural at a safe distance, tucked far away in time and space.[8] Michael Heiser rightly notes…

> …many Christians claim to believe in the supernatural but think (and live) like skeptics. We find talk of the supernatural world uncomfortable.[9]

It seems ironic that one could believe an almighty Creator spoke into nothingness and an entire universe sprang into existence, then doubt that the same Creator can create new limbs on a paraplegic. Can a Creator who designed the DNA of a vine, then swirled a solar system so that fruits grow in the sunshine of a balanced ecosystem, not be able to change molecules of water into molecules of wine? When God chooses to intervene miraculously, it's neither harder nor easier for him than creating the cause-and-effect universe in which he intervenes. To the motion of the smallest subatomic particle, he remains sovereign over his creation.

Our problems with miracles are directly proportional to problems with our view of God himself. As J.B. Phillips chided, "Your God is too small!"[10] Yet this is occurring despite the most intimate indication of the supernatural. Within our very souls—our inner selves—we feel hopelessly entangled with the transcendent. We sense that the Story *must* have a sequel beyond the front and back covers of the present plot.

I don't follow the naturalists' line that this experience is merely the psychological result of some evolutionary survival

---

[8] Another segment of contemporary Christianity pushes the view of the supernatural to the other extreme. In growing numbers, these believers pursue transcendent experiences as a regular component of worship. There is certainly nothing wrong with the Spirit-led desire for the miraculous, but not at the expense of balanced spiritual discipline. According to Heiser this group is "so busy seeking some interaction with [the supernatural] that it has become unconcerned with its biblical moorings…" In the end, contrived or exaggerated "miracles" lead away from scriptural truth as effectively as naturalism.

[9] Michael Heiser, *The Unseen Realm*, p. 17.

[10] J.B. Phillips, *Your God Is Too Small*, (New York: Simon and Shuster, 1952).

trait. Scripture informs us that God has "set eternity in the human heart."[11] The thought of an absolute "final page" is foreign to our spiritual DNA. We can understand a change from *this* to *that*, or *here* to *there*, but not *something* to *nothing*. One of the characters in the play *Our Town* sums it up…

> We all know that something is eternal. And it ain't houses and it ain't names, and it ain't earth, and it ain't even the stars… everybody knows in their bones that something is eternal, and that something has to do with human beings.[12]

And certainly not to be overlooked, true believers have the very Spirit of God living within. In some incomprehensible, yet palpable way, his Spirit communicates with ours. God graciously provides the gift of an immediate connection to *his* supernatural realm.

Roots of Doubt

Much of our supernatural apprehension can be traced to the emphasis on scientism that emerged during the Age of Enlightenment.[13] The shift from God-centered to man-centered paradigms in many different disciplines concentrated our thinking on material explanations rather than those unseen. Empirical thought became the preferred source of truth over mere faith and tradition…

> The liberalism of [the nineteenth century] denied all of the supernatural elements of the Christian faith,

[11] Ecclesiastes 3:11.
[12] Thornton Wilder, *Our Town: A Play in Three Acts,* (U.S.: HarperCollins,1938).
[13] As opposed to pure scientific methods and objective conclusions, scientism refers to the philosophy that advocates empiricism to the exclusion of other ways of knowing.

including the virgin birth of Jesus, His miracles, His atoning death, and His resurrection. The supernatural was stripped altogether from Christianity.[14]

It's doubtful that many would want to rewind all the changes brought by the Enlightenment. Antibiotics are my first choice for treating strep. Yes, I pray for healing, but I usually don't have to resort to exorcising a spirit of sickness. And good scientific reasoning has clearly revealed much about the creation. Many scientists eagerly used the scientific method as a tool of theology. At the dawn of this Age of Reason, Johannes Kepler saw this new empirical science as "thinking God's thought after him."[15] Nonetheless, the miraculous in large part has taken a backseat to the natural.

From 1985 to 2006, a group of Bible scholars and lay leaders met under the name *The Jesus Seminar*. Their objective was to determine which of the "Jesus accounts" in the Bible were true. The group of Christians essentially accomplished what Jefferson had done previously, namely, eliminating any associations of Jesus with the supernatural.

A more recent contributing factor has been the rise of the militant "new atheist religion" and their "deacons" within the naturalistic scientist congregation. A relatively small band of outspoken atheists has disproportionately muzzled many in academia, especially seminaries. Christian professors, scientists, pastors, and lay-thinkers, in general, are often bullied by the politically correct leanings of their peers. It's far easier to ignore biblical super-nature and dwell on social engagement.

To the dismay of leaders within radical atheism, most of those who identify as non-theist still seem to sense and believe

---

[14] R.C. Sproul, "Christianity is a Supernatural Faith," www.ligonier.org, Nov 11, 2013.

[15] Johannes Kepler, "A letter to the Bavarian chancellor Herwart von Hohenburg," 1599. Likely a paraphrase of "Those laws [of science] are within the grasp of the human mind; God wanted us to recognize them by creating us after his own image so that we could share in his own thoughts."

in something beyond.[16] But even though belief in the transcendent is widespread, a significant problem remains. Without a genuine relationship with the supernatural Christ of Scripture, unregenerate humans tend to seek some other path of transcendence, usually one without that old-fashioned sin-and-accountability stigma.

And adding to the confusion, many within mainline churches follow suit. They feel that the transcendent is undeniable, but distorted and human-oriented experiences seem to offer a better and easier way to pursue the supernatural.

One recent study revealed that belief in the biblical supernatural has no bearing on a person's belief in the non-biblical paranormal.[17] In other words, the average American could believe in God *and* ghosts with no notable contradictions. Other polls regarding ghosts, UFOs, and the paranormal in general, reveal that there isn't even a significant difference between believers and non-believers. Why do many Christians seem so hesitant to accept biblical supernatural events, yet dive headlong into the deception *du jour*? Both extremes undermine a truly Christian worldview.

How Many Miracles Do We Need?

If the truth about the supernatural is so important, why doesn't God accommodate? Wouldn't a few simple miracles here and there fill the churches? Why does it seem to be God's will to veil the supernatural? This "hiddenness" of God and the supernatural, in general, has long been a topic of debate among theologians. Although Scripture clearly assumes the veracity of the spirit realm, it gives but fleeting glimpses.

---

[16] John Templeton Foundation, "Towards a Psychology and Sociology of Atheism and Non-belief," www.templeton.org, 2019.
[17] Tom W. Rice, "Believe It or Not: Religious and Other Paranormal Beliefs in the United States," *Journal for the Scientific Study of Religion,* 2003, 42(1), 95-106.

We might also ask whether God is obliged to reveal himself through supernatural intervention at all. It's clear in Scripture that faith is an important element of knowledge. In Jesus's story about the rich man and Lazarus, the rich man basically asked that his brothers might receive supernatural insight regarding the afterlife. Without mincing words, Father Abraham tells him "If they do not listen to Moses and the Prophets, they will not be convinced even if someone rises from the dead."[18] Thus, faith in God's Word should be sufficient.

Perhaps as Pascal says, faith is God's priority rather than sight: "In faith, there is enough light for those who want to believe and enough shadows to blind those who don't."[19] Michael O'Neill states it well…

> Although seeking miracles is often an attempt to quench an authentic thirst for the spiritual…, miraculous phenomena are not a substitute for absolute faith in God.[20]

I believe that recovering a *balanced* scriptural view of the supernatural should be a high priority for the 21st-century Church. The supernatural realm and its interaction with this world are foundational truths. Miracles have purposes beyond their immediate results. Even a cursory study of the supernatural events in both the Old and New Testaments show that there are no gratuitous miracles. All fit exquisitely within the narrative and serve to glorify God, encourage believers, and silence doubters. The Christian worldview must include an accurate doctrine of supernatural reality in order to maintain its relevance and effectiveness in a deceived world.

---

[18] Luke 16:19

[19] Blaise Pascal, *Pensees.*

[20] Michael O'Neill, *Exploring the Miraculous,* (Huntington, IN: Our Sunday Visitor, Inc., 2015).

3
Supernatural Intrusions

"The Christian story is precisely the story of one grand miracle ... what is beyond all space and time, which is uncreated, eternal, came into Nature, into human nature, descended into His own universe, and rose again, bringing Nature up with Him... If you take that away, there is nothing specifically Christian left." C.S. Lewis[21]

"By faith we understand that the universe was created by the word of God, so that what is seen was not made out of things that are visible."
Hebrews 11:3

As they rush to the ER at midnight, a father and mother pray for their son to be healed from a serious case of pneumonia. The medical team goes to work. All the modern treatments are employed. Within hours, the child's breathing has improved. By the next morning, the doctor says the boy is out of danger. He is expected to recover fully.

---

[21] C.S. Lewis, "Miracles," *The Complete C.S. Lewis Signature Classics*, (New York: Harper Collins, 2002).

Did God work a miracle? Did prayer change the course of the physical world? Did God "use" the treatments to heal the child? Our pat, Christian answer is *yes, all of these...*

What does it take to qualify as a miracle? Our English use of the word has become so diluted as to be practically meaningless. It was a miracle that she passed the exam... He miraculously made the goal at the last second... The beautiful newborn is such a miracle... The cancer is gone! What a miracle! But how often does God directly (supernaturally) intervene in our physical realm? Let's narrow our definition of a miracle.

Rare events, in general, aren't miracles. It's highly unlikely that a particular person will win a lottery. Yet in most lotteries, at least *one* person *will* win. If you win, it will appear to be a miracle from your perspective. But from the perspective of the rest of us, you just happened to be *the one*. Random events may appear to be miraculous even though they can be explained with normal statistics. Let's consider these mere statistical occurrences *non*-miraculous.

For our purposes, let's also exclude what I'll call *providential miracles*. In a sense, everything that happens is the result of God's intervention, since he sovereignly chose to *intervene* and bring about the entire system that is the universe...

> Many people will talk about the miracle of a sunrise or of childbirth. Such references are generally well-intentioned, and individuals who say such things are certainly correct that God—because of divine providence—is involved in every sunrise and childbirth. But in the normal course of things, a sunrise or childbirth is not a miracle. God established the regular movements of planets and the process of human conception and childbirth, and any sunrise or birth occurring within such parameters is the result of

> God's ordinary providential working through means, not the extraordinary providence of miracles.[22]

I would classify most answered prayers as providential miracles. God *could* (and sometimes does) intervene supernaturally to heal a sick child instantly, but he usually works through normal physical means such as medication or the body's own defenses. These cases, though not directly miraculous, are no less attributable to God and we should give him equal glory.

The most direct miracles, the ones we think of as truly miraculous, are qualitatively different. I would define a true miracle as *an event in which a supernatural agent suspends at least one physical cause in bringing about a particular effect*.

A miracle then has a direct, supernatural source. As creatures of imagination and adventure, extraordinary occurrences often lead us to otherworldly speculations. For the term *supernatural* Webster's gives us "…relating to an order of existence beyond the visible observable universe." As a scientist, I might reframe this as *anything beyond that which my senses can detect or measure using the tools of science.*

Paranormal literally means "alongside the normal." It's commonly used interchangeably with the term *supernatural*, but in our discussion, let's define it more literally. I consider an event paranormal if its cause isn't explained by current science. Note the subtle difference when compared with the definition of *supernatural*. Paranormal allows for the possibility that the explanation *could* be caused by physical laws, but that not all physical laws might currently be known. In other words, a paranormal event might be supernatural or just *currently beyond* natural explanation.

---

[22] Tabletalk Magazine "Providential Working Through Miracles," www.ligonier.org, March 2017.

Our "Closed" Universe?

C. S. Lewis provides an interesting metaphor in his book *Miracles.*[23] He uses a fishbowl and bombs (as he was writing in wartime England) to illustrate his view of supernatural miracles. Allow me to update his model somewhat...

Recently, I happened to be watching my own aquarium at the very moment one of our minor earthquakes occurred. The jolt rocked the house only slightly, but the fish reacted violently, darting wildly about the tank. Now for all practical purposes, the aquarium was the fishes' universe. Glass walls bounded a limited space in which they could do whatever fish wanted to do, relative to the physical laws in that space. The Scientist Fish among them had access to a lot of data, but he was limited by the nature of his "universe." He had probably formed a more-or-less coherent set of hypotheses and laws based on the available data within the tank. The other fish came to him for explanations.

But Philosopher Fish might question some of fishdom's beliefs. What about the food that seems to magically appear in the upper waters each day? Explain that! And what about those bubbles? Yes, they come from the big tube, but where does that come from? And, hey, look at that glow coming from one of the solid walls. Sometimes it looks like vague shapes are moving beyond the wall.

Philosopher Fish might provide interesting talk, but most of fishdom didn't dwell on her concerns. A normal fish day would consist of events traceable to previous events within their physical universe. One fish might venture into another's territory and the intruder would be nipped. A shy fish could find a plant to hide behind. Bottom-dwellers would plow the gravel and successfully find food fragments. Life as usual.

---

[23] C.S. Lewis, "Miracles," *The Complete C.S. Lewis Signature Classics*, (New York: Harper Collins, 2002).

But the earthquake movement… That couldn't be deduced from the laws and objects within the aquarium. Philosopher Fish would ponder, Science Fish would hypothesize, yet both would be hard-pressed to come up with a good natural explanation. Rumors of a miracle would begin to circulate, and indeed, the event did fit the definition of a miracle.

So here I sit, a being in the "super-nature" outside of the fishdom universe, with a virtually complete understanding of the event. If I could communicate with the fish, I could clear up so many mysteries: *Your nature is just a smaller part of a much bigger nature! Things outside your world can and do affect your world! As long as you are fish in your fish realm, there will always be things you can't know for certain!*

From my perspective, there was no supernatural event. I had the ability to observe what Lewis called "the whole show." The brief turmoil in the aquarium was the result of a force beyond the fish universe, but it was no less causally related to a chain of real events than any of the day-to-day events within the tank. The only difference was that part of the chain was in another realm.

So, is it at least logically possible that our detectable universe is within a larger reality? This isn't a new idea by any means. Old and New Testament writers assumed such an arrangement, and since the mid-1900s, even the staunchest naturalists have become open to the idea. Many astrophysicists propose that the Big Bang had a trigger that was outside of the observable universe. Whether in mathematics or physical reality, string theorists also must consider several dimensions of time-space that are somehow outside of or beyond our dimension.[24]

---

[24] Barton Zwiebach, *A First Course in String Theory,* (Cambridge University Press, 2009).

The Gap

Yet even with full acceptance of realms beyond, there is still some uncertainty in designating an event as a miracle. As recently as the 1800s, most people believed the causal agent of disease was something supernatural. Viruses were far beyond detection by even the best science, but we now know that viral diseases are the result of very natural physical circumstances. Can we scream "Witch!" whenever a phenomenon is for the moment unexplained?

This has been called the *god-of-the-gaps fallacy*. Originally, the idea was a criticism by Christian thinkers aimed at "sloppy" Christian reasoning. It refers to the habit of wedging supernatural explanations into our gaps in knowledge. A classic example is Sir Isaac Newton's appraisal of solar system motions. The brilliant and devout scientist had carefully laid out the motions of the planets and comets with incredible precision. However, he also noted that gravitational forces among these bodies would tend to deteriorate these perfectly tuned motions. How can the consistent, steady orbits be explained with so many variables bent on messing them up? Just apply God's miraculous interventions as needed…

> This most beautiful System of the Sun, Planets, and Comets, could only proceed from the counsel and dominion of an intelligent and powerful Being.

Newton was correctly convinced of the divinity that brought about the elegance of the cosmos, yet he believed the Creator had to constantly step in and "fix" the gravitational errors that kept popping up. Later in the 18th century, Pierre Laplace refined Newton's calculations and found that natural causes *could* account for the maintenance of planetary orbits.[25] So, did this remove God from the picture? Does it follow that

---

[25] Thus, Newton's initial view might have been considered *paranormal* at the time.

all unknowns will eventually be explained by natural processes?

Clearly, discovering natural explanations for some knowledge gaps doesn't prove a totally naturalistic universe. Even if every single link in the chain of events within our "aquarium" were to be derived, would that rule out the supernatural? The most austere Deists might believe that the universe operates like a complex machine, yet they still postulate a supernatural origin for the machine. Most modern theories of cosmology, specifically the origin of the cosmos, are *ad hoc* ideas that attempt to explain the complex, low-probability nature of our reality. Our "fine-tuned" universe appears to be too "purposeful" to have been brought about by anything other than an intelligent, powerful (supernatural?) source.

Even though the Bible is absolutely sufficient in revealing God and his word, God doesn't limit us to this *special revelation.*[26] Paul tells us that God also reveals his supernatural character through *general revelati*on as we observe his creation.

> For his invisible attributes, namely, his eternal power and divine nature, have been clearly perceived, ever since the creation of the world, in the things that have been made. So they are without excuse. (Romans 1:20).

Nevertheless, *God does operate in some of the gaps...* We may not always know where this occurs, but we are assured that our physical realm will continue to provide evidence of the supernatural.

---

[26] Theologians often use the terms *special revelation* and *general revelation*.

All in the Numbers?

Hume and other philosophers have argued that miracles are impossible to identify because of our limited knowledge.[27] He claims we can never know whether an unlikely event (miracle) is due to supernatural intervention or the laws of nature and probability.

In one sense, this is true. With the previous lottery example, even if we decreased the chance of winning to one in "the number of sand grains on earth," there remains a statistical chance of winning by sheer natural laws. Thus, in theory, we can never *prove* empirically that an event had a supernatural cause. Lewis gave us a valuable construct for this uncertainty. He suggested we could address the problem by considering the "innate sense of the fitness of things" regarding the alleged miracle.[28]

Suppose it's a hot summer day and your car dies on a deserted stretch of highway. You have a baby on board who is becoming restless and did I mention—you forgot the bag with all the baby paraphernalia! What will you do? Suddenly, a baby bag falls onto the road beside you. Unknown to you, a baggage compartment door on a passing plane had come loose, allowing a traveler's bag to fall. The bag has diapers, water, and even toys! Then all at once, the sun begins to darken. Oh, you forgot that there would be a solar eclipse today? A cool shadow envelops the area as you give your baby a drink and change her diaper. As the eclipse ends and the baby sleeps, a large bus pulls up. It's your college roommate who is now a famous rock star. She invites you into the bus and agrees to take you home.

Wow! What a coincidence!

Really? Is it possible for the bag to fall as described? Well, yes. There is a painfully small chance, but it could certainly have happened by natural means. And the eclipse? Eclipses are

---

[27] David Hume, *An Enquiry Concerning Human Understanding,* Section X, 1748.
[28] C.S. Lewis, "Miracles," *The Complete C.S. Lewis Signature Classics*, (New York: Harper Collins, 2002), 396.

rare in any one particular place, but they happen *somewhere* with some regularity. It does seem like a big coincidence that your rock star friend just happened by, but that too is within the realm of natural occurrences.

Even if we consider the probability of all these events coinciding at a particular place and time (an *astoundingly* small probability!), we still must admit that statistically speaking, there is a chance. But let's face it—this was a miracle! And the reason we can (in faith) assume that is because of the "fitness" of the events. It isn't simply statistical reasoning that defines the miracle. The events occurred in a *context*—the context of your specific need. We intuitively know from the context, when a miracle has occurred. Skeptics might still balk, but appeals to purely statistical analyses fall by the wayside when viewed from a specific, unique narrative. Lewis explains…

> …miracles must, of course, interrupt the usual course of Nature; but if they are real they must, in the very act of so doing, assert all the more the unity and self-consistency of total reality at some deeper level. They will not be like unmetrical lumps of prose breaking the unity of the poem; they will be like that crowning audacity which, though it may be paralleled nowhere else in the poem, yet coming just where it does, and effecting just what it effects, is (to those who understand) the supreme revelation of the unity in the poem's conception.[29]

The story of Balaam's talking donkey (Numbers 22) is often cited as an example of the ridiculous nature of some biblical miracles. I must admit that this account did bother me at one time. Superficially, it seems somewhat on the silly side.

---

[29] C.S. Lewis, "Miracles," *The Complete C.S. Lewis Signature Classics*, (New York: Harper Collins, 2002), 355.

Why would God make a donkey speak when he has the power to do any number of more convincing miracles? But God's Word is hardly meant to be taken superficially. The event *in its context* is rich with meaning.

Balaam was a pagan diviner through whom God was about to give a prophetic message to his people Israel. Can God use a pagan sorcerer as a mouthpiece? Would his message be true, comprehensible, and valid? For an answer, we need look no further than the miracle. Could God use a donkey to give a message? Apparently so. In case Israel, contemporary readers, or even Balaam himself ever doubted God's ability to speak truth through whomever he pleases, the case is now closed. Had God willed, the donkey could have just as easily proclaimed the message instead of Balaam!

Our previous definition of a miracle was from God's view. Let's now describe miracles and supernatural events from our limited perspective: *An event with exceedingly low probability, which occurs meaningfully within a specific context.*

4
Paranormal Pickles

"All that glisters is not gold—
Often have you heard that told.
Many a man his life hath sold
But my outside to behold."
William Shakespeare

"The skeptics have to be right every day, but the believers only have to be right once."
Nick Pope[30]

Modern information collecting has created a monster for paranormal researchers. The data are vast for most paranormal phenomena. For example, one UFO research group claims to have investigated 120,000 cases over the past fifty years[31]. When faced with the mountains of information the squeamish usually shelve the topic. Thus, my cucumber metaphor…

Years ago, my family grew cucumbers for a large pickle company. It was a small patch, but it kept us quite busy through much of the summer. For those unfamiliar with the art of cucumber cultivation, the pre-pickle fruit is the same color as the leafy vines, leaving them well-camouflaged during harvest.

---

[30] Nick Pope, former UK MoD Investigator, www.nickpope.com.
[31] Ben Brazil "Irvine-based UFO Group Marks 50 Years of Watching the Skies," www.latimes.com, June 26, 2019.

As hard as one might try to pick each cucumber at its peak (smallish, straight, and succulent), each pickle-picking reveals numerous "escapees" from previous days.

Cucumbers grow very quickly in the summer heat, and a fingerling missed one day could be almost watermelon-size two days later. It was amazing how many humongous or distorted fruits appeared in the daily harvest. The "monsters" had no market value, so they became chicken feed. Only a small proportion of the harvest was the small, quintessential, *pickles de résistance* that the picklers wanted. We watched anxiously as the sorting machine culled out the different sizes and shapes into the appropriate bins. Sadly, the pickup pile of potential pickles typically produced only a paltry portion of premium pickles for profit. Hence the *Pickle Principle…*[32]

As previously mentioned, most alleged paranormal data are very large collections of *everything*. And by *everything,* I mean *everything…* Information that ranges from stupid to plausible… Obvious to unexplainable… Personal to widespread… Anecdotal to multi-documented… This is a major reason for the broad range of opinions regarding the phenomenal. If someone wants to show that ghosts are hoaxes, there are plenty of data points that will support that hypothesis. If another claims that only insane people see ghosts, there's a pile of good evidence for that as well. For the skeptic, the scientist, or the true seeker, the "premium" paranormal scenario would be 1) well-documented, 2) from a reliable source, 3) researchable to some extent, and 4) in opposition to at least one physical law. Yet keep in mind, as much as we would like to find our pickles neatly sorted and canned—it won't happen. *If* the authentic exists at all, it will be buried in a truckload of distracting "problematic pickles."

To be clear, I'm not suggesting that a small number of genuine supernatural phenomena *will* be hiding in every data set. I could be wrong, but I'll go out on a limb and say that out

[32] Similar to the composition fallacy.

of all the cat owners who say their cat speaks to them in whole sentences, *not one of them* is experiencing something paranormal. (There's probably a more psychological explanation.) *All the available data might support natural explanations.*

For our purposes, I propose three general classifications for miracles and other paranormal phenomena:

1. Mistaken identification.
2. Directed human activity.
3. Observationally accurate event.

Humans are quite prone to make *mistaken identifications*, especially in the midst of an unusual situation. Our brains purposely focus on some stimuli and ignore others. Part of the problem in discerning the paranormal is that it essentially is "alongside" the normal. It isn't totally *opposite* or *un*-normal. There may be at least some element of reality associated with it, thus, something natural may be mistaken for something unnatural.

Directed human activity includes deliberate deception for a variety of reasons such as profit, power, malice, entertainment, or national security. We certainly don't have to look far for examples of trickery and collusion being passed off as paranormal.

Observationally accurate implies that the phenomenon isn't explained by one of the first two categories, although a supernatural cause isn't *necessarily* being asserted. For example, in the case of a supposed miracle, this simply means that the observation corresponds in reality to the perception: "I saw a paraplegic person take up his bed and walk." The observer actually observed (not necessarily interpreted) accurately. In a poltergeist account, the report may say: "A cup levitated over the tabletop for several seconds then dropped and broke." Although this would be a very odd incident, loaded with all kinds of paranormal hypotheses, the observation itself doesn't cloak a cause or motive. Often, phenomena are reassigned as more information is gathered. For instance, what appears at first to be observationally accurate may be determined a mistaken identification or vice versa.

The PICKLE PRINCIPLE

In a collection of alleged paranormal data, most or all the incidents can be explained by natural means, however, that doesn't preclude the fact that some incidents in the collection could be paranormal.

5
The Ephesians Hypothesis

"For we do not wrestle against flesh and blood, but against the rulers, against the authorities, against the cosmic powers over this present darkness, against the spiritual forces of evil in the heavenly places."
Ephesians 6:12

Rocket science must be complicated because I'm constantly reminded of things that *aren't* rocket science. It's complicated because there are so many variables that could potentially affect a rocket's flight. What if a rocket scientist forgot to take into account an important variable such as gravity? That would be disastrous.

Even a surface reading of Scripture shows that the unseen realm is a significant variable in earthly events. We are biased toward the physical realm, though. It's what slaps us in the face every morning and dominates our thoughts until our eyes close at night. The physical is *here and now*, the spiritual seems *somewhere and sometime*. But the Bible paints a spiritual realm that is just as real (if not *more* real) than the physical. When enemy troops were attacking and hope was running low, Elisha could see a reality that was invisible to his servant…

> He said, "Do not be afraid, for those who are with us are more than those who are with them." Then Elisha prayed and said, "O LORD, please open his eyes that he may see." So the LORD opened the eyes of the young man, and he saw, and behold, the mountain was full of horses and chariots of fire all around Elisha (2 Kings 6:16-17).

Suddenly, the servant could see the vast supernatural army that was standing by, ready to intervene in the earthly battle. In the New Testament, Paul speaks of the same reality. At the end of a list of very practical, "earthly" teachings, he adds another item…

> Finally, be strong in the Lord and in the strength of his might. Put on the whole armor of God, that you may be able to stand against the schemes of the devil. For we do not wrestle against flesh and blood, but against the rulers, against the authorities, against the cosmic powers over this present darkness, against the spiritual forces of evil in the heavenly places (Ephesians 6:10-12).

Paul doesn't switch thoughts here. In the apostle's mind, a typical day involves visible, earthly activities intertwined with the invisible supernatural. He indicates that God expects us to view the physical world in the context of the larger spiritual realm. The terms "armor," "schemes," and "wrestle" imply serious business. A battle rages whether we choose to acknowledge it or not. Moreover, the unseen is actually of greater concern since the true struggle is *not* with the "flesh and blood" of this world. Observable, physical events sometimes come from "darkness," "cosmic powers," and "spiritual forces of evil."

Is this too disturbing? I was confronted by a good Christian lady once who told me "The Christian life is supposed to be peaceful and joyous. From what you're saying, I would be scared to get out of bed every morning!"

It's unfortunately true that the idea of demonic spirits sneaking around, spreading mischief in the physical realm is ridiculed by liberal theology. Many would prefer to leave these images on the ceiling of some medieval cathedral. There's no doubt that the popular message from most pulpits is strictly one of peace and joy. I don't deny that these are the ultimate fruits of the Spirit's work in believers' lives. Yet the Bible demands that an understanding of reality *must* include an understanding of the realm beyond our earthly perceptions. The formula must address all the variables. Our worldview is incomplete when data are limited to the material world, and the unseen enemy has free rein to engage us.

Let me be very clear before going further. I do NOT believe there is a demon behind every bush (although there is very likely a bush in front of every demon). I have no intention of sensationalizing the "dark side" in a paranoid witch-hunt. Even though the struggle is real, the outcome is already decided. The command to "be strong in the Lord" assumes that we *can* win. *If* we "put on the full armor of God," we *will* "stand against the devil's schemes." Paul's words are as much encouragement as they are warning.

Living with the Unseen

In the 19th century, Louis Pasteur had the idea that some diseases might be caused by particles too small to see. Many questioned this bizarre concept and refused to accept an unseen world. Pasteur reasoned that there should be a way to test this *hypothesis*. *If* these particles did exist, and *if* they did cause disease, and *if* they could be passed to others, he should be able to predict certain scenarios. For example, a person who had kissed a sick person would get the disease, but a person who stayed far away wouldn't.

A dictionary definition of *hypothesis* is "a tentative assumption made in order to draw out and test its logical consequences."[33] I think of a hypothesis as an educated-guess-starting-point for investigating a question or idea. A researcher usually looks at a collection of available information and proposes a possible scenario that could explain it. *Testing* (validating) the hypothesis requires comparing real-world observations to the proposed scenario.

The set of propositions Paul presents in Ephesians 6 forms a very useful hypothesis for examining phenomena that may be connected to paranormal or supernatural beings. Consider the specific truths he offers…

- There is an unseen realm beyond the one in which we live.
- Spirits from that unseen realm are able to interact with our world.
- Our primary struggle is with fallen spirits from this unseen realm.
- Fallen spirits are evil, scheming, and set on doing us harm.
- We can and should be prepared to recognize and withstand these fallen spirits.

Assuming these propositions are true, we would expect this *Ephesians Hypothesis* to be testable in our physical world. In other words, we should be able to make predictions about how the unseen interacts with the seen—how the supernatural reacts with the natural. And, if the Ephesians Hypothesis is correct, we would expect it to have explanatory power with

---

[33] merriam-webster.com

some phenomena, especially those that appear to have causes beyond the physical universe.

Again, this isn't to say that every unusual or hard-to-explain phenomenon is supernatural. However, I do believe that *there is only one supernatural reality, as described in Scripture*. Therefore, if a supernatural cause is indicated, *it will necessarily be of the type described in Ephesians 6*—not some contrived reality, made to fit a particular phenomenon. Below are the full implications of the Ephesians Hypothesis. We'll refer to these often. In the following chapter, let's take a closer look at what Scripture reveals about this unseen realm.

Expanded Ephesians Hypothesis

> For we do not wrestle against flesh and blood, but against the rulers, against the authorities, against the cosmic powers over this present darkness, against the spiritual forces of evil in the heavenly places (Ephesians 6:12).

Presupposition: If a supernatural cause is indicated, it will necessarily be of the type described in Ephesians 6 and nothing else.

Ephesians 6 Principles

1. There is an unseen realm beyond the one in which we live.
2. Spirits from that unseen realm are able to interact with our world.
3. Our primary struggle is with fallen spirits from the unseen realm.

4. Fallen spirits are evil, scheming, and set on doing us harm.
5. We can and should be prepared to recognize and withstand these fallen spirits.

Other Scriptural Truths Supporting the Ephesians Hypothesis

a) There is only one supernatural reality—the unseen spirit realm described in Scripture. If the supernatural is indicated, it will be of this type, and this type only. (John 14:6)
b) Paranormal events will never be neutral. They are designed to lead the observer. (1 Peter 5:8)
c) The inhabitants of the unseen spirit realm include the eternal, self-existent, triune God and the spirit beings he created. (Job 1:6)
d) Biblical spirit beings can interact with the physical world. (Colossians 2:8, 1 Corinthians 10:20, 1 Timothy 4:1, Daniel 10:13, 2 Corinthians 4:4, 1 Thessalonians 2:8, 2 Corinthians 2:11)
e) Some of these spirits, including one of great power, rebelled against God. (Ezekiel 28:11-19, Isaiah 14:12-19)
f) Fallen spirits are able to "lead" the thoughts of humans, as allowed by God. (2 Corinthians 2:10-11, Ephesians 4:26-27, 1 Timothy 3:6, 1 Corinthians 7:10, Acts 5:3, 1 Peter 5:6-10)
g) Not all evil is directly caused by fallen spirits. Its source can be our own fleshly nature and the evil of the world system. (Ephesians 2:1-3)
h) Rebellious, fallen spirits are able to possess and control non-believers, as well as animals. (Matthew 8)

i) Satan, the leader of fallen spirits, has the goal of opposing God in any way possible. (2 Corinthians 2:8, 1 Corinthians 10:20, Zechariah 3:1, John 8:44, Matthew 4:3)

j) Satan's chief strategies are lies and deception. (John 8:44, 2 Corinthians 11:3)

k) Satan and fallen spirits will use any means possible to deceive humans and lead them away from God and truth. In most cases, it's in Satan's interest to not reveal his true identity. Demons will manifest themselves to some, but maybe not others. (2 Corinthians 2:11, 1 Peter 5:8)

l) Satan can appear as an "angel of light." (2 Corinthians 11:14)

m) Fallen spirits have some ability to manipulate matter and energy. (Revelation 9, Job 1, Mark 4:39)

n) Although Satan and fallen spirits have great power, the ultimate power is God. (1 John 4:4, Revelation 10:20)

o) God in his love and omniscience often allows a temporary *lesser evil* in order to gain an eternal *greater good*. For example, in revealing sin, leading to repentance, or sanctifying believers. (Romans 8:28, 2 Corinthians 4:17, Book of Job)

6
Dwellers in the Supernatural Realm

"Human history is seen to be not merely an account of human activities and events independent of spiritual forces, but a continuous interaction of spiritual and human personalities, in which demons play a prominent part." Merrill F. Unger[34]

Before examining the alleged supernatural, let's consider the supernatural realities presented in Scripture. The Old and New Testaments give brief but important glimpses of the divine entities who inhabit heaven with the triune God. These beings are often referred to as the "sons of God," (Hebrew, *beney Elohim*) a term typically used to describe direct creations of God. Sometimes the Old Testament designation for these creatures is simply *Elohim*, translated "god" or "gods" (depending on the context, it may be singular or plural).

Elohim is used over a thousand times to refer to the one God of Israel, but it can denote any lesser spirit being or group of spirit beings. The confusion in English translations is that *Elohim* isn't necessarily a name, or even a reference to a specific set of attributes (omniscience, omnipotent, self-existent).[35] Rather, it speaks more of a being's domain, and

---

[34] Merrill F. Unger, *Biblical Demonology: A Study of Spiritual Forces at Work Today*, (Grand Rapids, MI: Kregel, 1994), 181.

[35] Context tells when *Elohim* refers to "The God". He is also called by his true name, YHWH (Yahweh, Jehovah, I AM) or *Elyon* (Almighty God). For an in-depth treatment see *The Unseen Realm*, by Michael Heiser.

could be rendered "divine ones," "heavenly beings," or "dwellers in the supernatural realm." It would be equivalent to calling me a "southerner."

The Most High

"No one is like you, LORD; you are great, and your name is mighty in power."[36] Sadly, even in Christian circles, this truth tends to become cliché. The words are so easy to say, yet the truth of them is unfathomable. I hesitate in my utter incompetence to write about God Almighty. How could a library of books begin to describe the unbounded Creator of all that is, much less a few paragraphs? The short answer—it can't. A.W. Tozer writes that the human mind will never "comprehend" God, but it can spend eternity "apprehending" him.[37] In this short section, we'll limit our focus to truths about God that help us understand certain topics in this book.

Rather than a proof of God's existence, the Bible begins with an affirmation: "In the beginning, God…" The Apostle Paul states that God's handiwork around us is sufficient evidence for his existence. He further claims that the universe tells something of the Creator's nature….

> For his invisible attributes, namely, his eternal power and divine nature, have been clearly perceived, ever since the creation of the world, in the things that have been made. So they are without excuse (Romans 1:20).

---

[36] Jeremiah 10:6

[37] For an excellent introduction to God's attributes see A.W. Tozer's *The Knowledge of the Holy*, (New York: Harper Collins, 1961).

But God desired a closer relationship than mere acknowledgment. We know some of God's other *attributes* (truths about him) only because he chose to *reveal* them to us through his Word…

> God the Creator is transcendent, mysterious, and inscrutable, beyond the range of any imagining or philosophical guesswork of which we are capable—and hence a summons to us to humble ourselves, to listen and learn of him, and to let him teach us what he is like and how we should think of him.[38]

Thus, any characteristic we assign to God from our own reasoning or wishful thinking is tantamount to idolatry since we're essentially "creating" a god in *our* image. Thankfully, through the Old and New Testaments, God progressively revealed more and more of his nature. He is…

- Holy – Completely separate from and intrinsically unlike all else (1 Samuel 2:2).[39]
- Triune – In a way we cannot comprehend, God is three persons in one essence. He revealed himself as Father, Son, and Spirit (Philippians 1:2, Titus 2:13, Acts 5:3–4).
- Personal – God is sentient, emotional, volitional—a being, as opposed to simply an ethereal, impersonal force (Exodus 3:14).
- Eternal – Without beginning or end (Hebrews 13:8).
- Unchanging – (immutable) God is always the same. Being perfect there is no need to change (Malachi 3:6).

---

[38] J.I. Packer, *Knowing God*, (Downers Grove, IL: InterVarsity Press, 1973), 48.
[39] Verses given are examples among many others. See Tozer for more.

- Self-sufficient – He needs nothing from any other source (Ephesians 3:20).
- Sovereign – He acts independently by his own nature to do as he pleases (Daniel 4:35).
- All-knowing – (omniscient) He knows everything that is or can be known (Isaiah 46:9-10).
- All-powerful – (omnipotent) He can do anything that is consistent with his other attributes (Jeremiah 32:17).
- Unbounded – (omnipresent) He is everywhere at all times (Psalm 139:7-12).
- Loving – He is perfectly devoted to the well-being of others (1 John 4:7-8).
- Righteous – There is nothing evil or imperfect in him (Psalm 71:19).
- Good – He is absolute goodness in all ways (Psalm 34:8).

In what has been called the Grand Miracle, the second member of the Trinity became a human like us. The New Testament tells of Jesus, the Son of God—God in flesh—who dwelled with us, then died as a sacrifice for our sins. Jesus—God *and* man—not only gave believers eternal life, he showed us in a physical, "this-realm" body something of what the unseen, "other-realm" God is like.

What Can't God Do?

It sounds very "churchish" to insist that God can do *anything*. He is, after all, omnipotent. We also describe him as unlimited, unbounded, absolute, and perfect.[40] But are there things that God simply can't do?

Immediately the old philosophy class argument comes to mind: *Can God create a stone so big that he can't lift it?* The question itself reveals faulty thinking about God and his attributes. Even the word *attribute* is a problem when applied to God. When referring to a human, we can think of these separate components of who we are. But God by his nature has no "parts" (not referring to trinitarian Persons here). He is simple unity, indivisible. The attributes he has revealed serve in a small way to describe him in terms our human minds might understand.

Being a perfect unity, any one of his perceived attributes can't negate or diminish any other attribute. For example, why can't he simply pardon our sins just because he wants to? Because that application of his *mercy* would contradict his attribute of *justness*.[41] God is also holy, or "set apart" from all his creation. This requires that he *not* be some things. He isn't equivalent to anything he has created. God can't be the same as something he created, so there are many things his nature forbids him from being.[42]

God is also unable to do something that can't be done. This may sound like a limitation or just plain silliness, but it's really only a language problem. The stone-lifting enigma is a logical impossibility. In other words, it isn't logically possible for "A" to be "not A", or for 1 + 1 to not equal 2.[43] In simplistic terms, logical contradictions and absurdities don't exist in reality, thus whether God can "do" them is a non-question. This doesn't mean that logic and mathematics are things that transcend God,

---

[40] I don't use the term infinite because I consider his attributes qualitative rather than quantitative. Infinity implies too many paradoxes and mathematical problems.
[41] Thus, Christ had to be sacrificed as an atonement.
[42] And his nature is not a separate force. It is *of* him. He is completely free from anything outside of himself.
[43] Some add "at the same time and in the same sense," but God is actually outside of these parameters.

but that logical consistency itself is based in the nature of God. Lewis states it very well…

> You may attribute miracles to Him, but not nonsense. This is no limit to His power. If you choose to say, "God can give a creature free will and at the same time withhold free will from it," you have not succeeded in saying anything about God: meaningless combinations of words do not suddenly acquire meaning simply because we prefix to them the two other words, "God can." It remains true that all things are possible with God: the intrinsic impossibilities are not things but nonentities… nonsense remains nonsense even when we talk it about God.[44]

Scripture gives several clear truths about God's limitations. Hebrews 6:18 tells us that it's "impossible for God to lie." For him to be untruthful would be in opposition to his character. A comprehensive study reveals other impossibilities for God, such as changing, being unfaithful, acting unjustly, and anything else contrary to his nature.

I emphasize God's limitations here because in dealing with paranormal phenomena in the following chapters, the question will frequently arise: "But *couldn't* God do it that way?" Our knee-jerk response is usually "Of course. God can do anything." However, our conclusion must be based on the reality of God's potential actions and consistent with his revealed nature.

Could God send an angel to tell me a lie? No, his nature wouldn't allow him to lie.

Could God punish me for a sin for which Christ has already atoned? No, God is completely just.

---

[44] C.S. Lewis, "The Problem of Pain," *The Complete C.S. Lewis Signature Classics*, (New York: Harper Collins, 2002), 561.

Could God change from the Old Testament to the New? No, God may deal differently by his choice, but he's perfectly unchanging.

In addition to considering God's *nature*, we must also take into account his typical *mode* of operation. Within the limitations of his nature, God is utterly sovereign…

> Whenever God acts, he acts in a way that pleases him... He is never backed into a corner where his only recourse is to do something he hates to do. He does whatever he pleases.[45]

At the heart of this, though, we find that even in two similar situations, God isn't bound to deal identically. He is completely free and just in punishing the sins of one nation immediately and delaying punishment for the same sins of another nation. He is equally free to bless you and not bless me, *simply because he chooses.*

However, God created a universe that is orderly, consistent, and for the most part predictable. Scientific reasoning depends on this *repeatability*. Likewise, he reveals to us through his Word that he is generally quite consistent in his interactions with us.[46] For example, most people who have ever lived have eventually died. Enoch, however, "… walked with God, and he was not, for God took him..."[47] God was sovereignly free to "break the rules" and allow his friend Enoch to bypass death. But I don't base my expectations on Enoch's uncommon experience. I expect to meet God by his usual mode. *Could* God take me, as he did Enoch? Of course! God is free to choose whatever he wishes, and there is nothing in his nature

---

[45] John Piper, "The Sovereignty of God," www.desiringgod.org, Accessed Feb 19, 2020.

[46] Some make a distinction between God's *principles* and his *promises*. He always act according to his promises. Principles are more general, applying most of the time, but not always.

[47] Genesis 5:22.

that would preclude it. *Yet, I must base my expectations on God's usual actions and not on the rarest of events.*

Could God send an angel to answer one of my questions? Yes, but it's much more likely that he will use an earthly method.

Could God perform a miracle to solve one of my problems? Yes, but he will probably use normal, everyday means.

Could God give me a new revelation through a dream or vision? Yes, but it's *extremely* likely that he will simply show me a truth already in his Word.

So, as we offer solutions for various phenomena, our first step is to determine whether that proposition is in any way contrary to God's *nature.* If it isn't, then the next step is to determine whether the proposition is consistent with God's *usual mode of operation… Could he? Would he?*

As humans observing unusual events, our tendency will almost always be to assume we are observing the exception to the rule. We see ourselves as the outliers, the anointed, and the chosen. As we'll find, that's a dangerous starting point for discerning the paranormal.

The COULD HE? WOULD HE? Test

When questioning whether God might be using supernatural intervention, first ask "Is this consistent with the nature of God?" If so, then ask "Is this typically the way God does it, or is there usually a more natural means?" You may be witnessing a miracle, but probably not…

Angels and Others

When we think of heavenly beings, our minds immediately bring up angels. The word *angel* in Scripture literally means *messenger*. Scripture sometimes uses the term for human messengers, as well. Like the term *Elohim*, "angel" doesn't refer to a set of attributes, rather it's a job description. We commonly use the term angel to refer to all heavenly beings (other than God himself), but that might be too simplistic.[48] The Bible is usually more specific in its terminology for heavenly dwellers. As mentioned previously, "sons of God," is the inclusive term for all the divine beings created by God. Angels are one type of "sons of God," or more accurately, *angel* (messenger) is a role for some of them.

Divine beings are mentioned often in Scripture. There is no section devoted exclusively to describing them, but many of their characteristics are diffused throughout the Bible. For example, we know that they were created by God before he created the earth and that they are innumerable. They are immortal, possessing personality, intelligence, and will. (Humans do *not* become angels when they die. Why can't Hollywood get that right?)

Heavenly dwellers are spirits, yet they may enter the physical realm in some kind of visible body or can appear in ordinary human form. Presently, they are superior to humans in ability and intellect. There is also a hierarchy within the heavenly ranks. Three divine creatures are named in the Bible—Michael, Gabriel, and Lucifer. Michael is known as an *archangel* (high angel), implying that he outranks others. Before his fall, Lucifer was once an *anointing cherub*, apparently another high-ranking being. Scripture mentions other unnamed cherubim (plural), as well as other heavenly beings, known as *seraphim*.

---

[48] Although in the OT, the "angel of the Lord" might be referring to the 2nd person of the Trinity.

Paul hints at other "types" or "job descriptions" of divine entities in Ephesians 6. He specifically refers to "rulers," "authorities," "powers of this dark world," and "spiritual forces of evil." We have no organizational charts given to map out the heavenly population or its hierarchy. It's strictly speculation, but I feel the heavenly realm is much more diverse and complex than we realize. God's creation of life on earth is so overwhelming I can only imagine that the eternal spirit realm is orders of magnitude more spectacular.

A Divine Council?

Many passages in the Old Testament have God speaking in plural pronouns—*we*, *our*, *us*. Some see the trinity in these references, which is certainly consistent with God's revealed nature. Others have suggested that this is a "majestic" use of plurals, as when the queen asks for her tea: "We'll be having our tea now." At times, though, members of this plurality are clearly distinguished from the Most High. Micaiah's vision depicts a scene in which Yahweh is allowing a "committee" of spirits to participate in the handling of an earthly incident…

> And Micaiah said, "Therefore hear the word of the LORD: I saw the LORD sitting on his throne, and all the host of heaven standing beside him on his right hand and on his left; and the LORD said, 'Who will entice Ahab, that he may go up and fall at Ramoth-Gilead?' And one said one thing, and another said another. Then a spirit came forward and stood before the LORD, saying, 'I will entice him.'" (1 Kings 22:19-21).

The *divine council* isn't a new idea. Jewish Old Testament tradition and early church leaders acknowledged a supernatural assembly with Yahweh (God's covenant name) as its head. As we have seen with other concepts, when the supernatural becomes *too* supernatural, a more "earthly" view is often substituted. Also, some take the idea of God conferring with other entities in heaven to be demeaning of the sovereign God or even polytheistic. The accusation is that it seems too much like Mormonism or Norse mythology.

However, Jewish and early Church teachings on the divine council never implied numerous self-existent beings, equal to or nearly equal to God. Unlike similar beliefs among their pagan neighbors, the Israelites were adamant about the absolute holiness and superiority of Yahweh above any of his created council members.

Why would God need a divine council? Couldn't he govern heaven alone? It's ultimately the same question as "Why does God need anything?" He doesn't… Yet in his perfect, sovereign will he delegates authority and assigns tasks to his creatures. It seems that in the same way God allows his earthly children to make decisions and carry out his plan, he similarly involves his heavenly population. [49]

The membership of Yahweh's divine council is debatable. Whether it consists of all the sons of God or only a select few is only conjecture. As previously mentioned, we do know of several roles filled by spirit beings such as angels, seraphim, cherubim, rulers, powers, thrones, and authorities.

---

[49] Lewis explains God's propensity to delegate: "He allows soils and weather and animals and the muscles, minds, and wills of men to co-operate in the execution of His will. 'God,' said Pascal, 'instituted prayer in order to lend His creatures the dignity of causality.' But not only prayer; whenever we act at all He lends us that dignity. It's not really stranger, nor less strange, that my prayers should affect the course of events than that my other actions should do so." He goes on to say that God's will hasn't changed, it has only been accomplished through the wills of his creations. C.S. Lewis, *The World's Last Night and Other Essays,* (New York: Harcourt, Harvest Books), 8-9.

The Rise and Fall of Lucifer

Originally, Lucifer was good, powerful, and flawless since God isn't the source of any evil or imperfection. His created beings, however, have the freedom to choose. In one of the most catastrophic and incomprehensible events in cosmic history, Lucifer, one of the "sons of God" used his freedom to oppose his Maker. Bible scholars see this passage as referring to Lucifer…

> You were the seal of perfection, full of wisdom and perfect in beauty… You were in Eden, the garden of God; every precious stone adorned you… You were anointed as a guardian cherub, for so I ordained you… You were on the holy mount of God; You were blameless in your ways from the day you were created till wickedness was found in you (Ezekiel 28:12-15).

The details are scattered, but Lucifer's primary sin seems to be pride. He set his mind to be "like the Most High…"

> How you have fallen from heaven, morning star,[50] son of the dawn!… You said in your heart, "I will ascend to the heavens; I will raise my throne above the stars of God; I will sit enthroned on the mount of assembly… I will ascend above the tops of the clouds; I will make myself like the Most High" (Isaiah 14:12-14).

Most theologians assume that this fallen entity corresponds to Satan (the *accuser*) and to the devil (literally,

---

[50] *Morning star* also translates as *lucifer* or *shining one*.

slanderer).[51] Jesus addressed this being as "Satan," thus, I'll follow his example.

Liberal theology portrays Satan as merely an impersonal "force of evil." Alarmingly, only 26% of Christians believe he even exists.[52] Satan is often viewed as merely the "personification of Evil" in collective humankind. Scripture, however, leaves absolutely no room for this notion. The Bible reveals him as personal, powerful, cunning, and single-mindedly bent on opposing Yahweh God by any and all means.

Some are surprised to learn that Satan actually does "rule" this physical world. In God's all-wise justice system, Adam legally relinquished dominion of the earth to Satan. This is how Satan could rightfully offer Jesus earthly authority over nations in the wilderness temptations. Indeed, one of the Deceiver's titles is "Prince of the Power of the Air."[53] Billy Graham aptly describes Satan's reign on Planet Earth…

> Since the fall of Lucifer, that angel of light and son of the morning, there has been no respite in the bitter Battle of the Ages. Night and day Lucifer, the master craftsman of the devices of darkness, labors to thwart God's plan of the ages. We can find inscribed on every page of human history the consequences of the evil brought to fruition by the powers of darkness with the devil in charge.[54]

However, be very clear about this… *Satan is not the opposite of God.* As great as he is in many respects, he has never been God's equal in any sense. Satan is a finite creation of the Most High. God is sovereign, self-existent perfection. Satan isn't omnipresent. He isn't omniscient. He isn't omnipotent. He

---

[51] The term translated *Satan* is literally "the satan" (the *accuser*) and is a "role" rather than a name.
[52] The Barna Group, Ltd, "Most American Christians Do Not Believe that Satan or the Holy Spirit Exist, Research Releases in Faith & Christianity," April 13, 2009.
[53] Ephesians 2:2.
[54] Billy Graham, *Angels: God's Secret Agents*, (Dallas: Word Publishing, 1975), 70.

operates only within the bounds that God has set.[55] Martin Luther called Satan "the ape of God." Yet in ways we cannot completely understand, God sometimes allows and uses the evil Satan wreaks to accomplish his own righteous ends. Despite his temporary rulership of the earth, our Enemy has already been judged by God. His destiny is set. At a point in the future, Satan will receive his final sentence and be banished forever. Jesus will then take his rightful throne as King of Kings.

Anytime we focus attention on Satan, even to arm ourselves against him, there is a risk—not specifically from him, but from our own inquisitive nature. Lewis warns…

> There are two equal and opposite errors into which our race can fall about the devils. One is to disbelieve in their existence. The other is to believe, and to feel an excessive and unhealthy interest in them. [Demons] are equally pleased by both errors and hail a materialist or a magician with the same delight.[56]

It's crucial that we as children of God understand our position in Christ. As we study paranormal phenomena and their potential connection to the Evil One, we should never overlook the promise we have been given…

> Little children, you are from God and have overcome them, for he who is in you is greater than he who is in the world (1 John 4:4).

---

[55] See Job 1-2

[56] C.S. Lewis. *The Screwtape Letters*, (San Francisco: Harper, 1942).

A Fallen but Formidable Army

Satan has an army of supernatural cohorts. There are several scriptural references to unrighteous "sons of God" who, like Satan, stand in opposition to the Most High. The exact origin of these evil *Elohim* isn't spelled out in detail. We can be sure God doesn't create evil, and there are no beings that he didn't create. Thus, any spirits who oppose God are "fallen" in the literal sense. Several passages indicate that Satan and possibly his followers fell from their states of righteousness before the creation of the earth (or at least humans). Some scholars believe that another group of beings may have fallen around the events of Genesis 6:1-4. Still another theory is that several other sons of God turned against the Most High sometime after Babel. Whatever the case, the practical results are the same: *A large number of evil entities now dwell in the unseen, supernatural realm.*

The term *demon* applies to at least some of these fallen beings.[57] As with the righteous angels, there is a hierarchy among the fallen ones. The highest level might include the "princes" referenced in Daniel 10. These high-ranking beings appear to be assigned to entire nations. Paul lists some of the other titles in Ephesians 6. The exact command structure isn't given, but as mentioned, these roles include "rulers," "authorities," "cosmic powers" in this dark world," and "spiritual forces of evil" in heavenly places.

If you have trouble accepting the literal, scriptural view of demons, you might tend to tune out at this point. Scripture, however, doesn't leave the option open to debate. Bible scholar Charles Ryrie says bluntly…

---

[57] Scholars suggest that demons could be 1) the disembodied spirits of a pre-Adamic race, 2) the offspring of unions between human woman and rebellious *elohim* (Gen. 6), or 3) angels who rebelled with Satan. Merrill Unger aptly assured that their origin is not as critical as the fact that "they are evil and harmful spirit personalities, that in their fellowship there is no safety, and that against them continual warfare must be waged."

> To deny the existence of demons is not skepticism; it only displays ignorance. To be unrealistic about their power is foolhardy.[58]

In attempts to point out the futility of following Satan, medieval artists set the stage for casting him as an idiot and buffoon. That might have worked with dark-age congregations, but today It's hard to take seriously an imp with a red suit, pointy tail, and pitchfork. I'm sure Satan doesn't mind the caricature. Is there a more effective way to work covertly than to be deemed unthreatening or nonexistent? It's even better when those who *do* believe in demons are labeled as illiterates and fanatics.

And clearly, Satan, demons, hell—all the negatives—are the lowest of priorities in liberal churches. These topics certainly put a damper on seeker-sensitive strategies. Yet the Bible makes no apologies for its portrayal of evil beings. Every New Testament writer warns against Satan and his lies. Unless you happen to be reading from the Jefferson Bible, why would Jesus spend so much time, effort, and prayer dealing with Satan and his demons if they were imaginary or unthreatening?

Spirits in the Earthly Realm

So, does this angelic realm (including righteous and fallen) contribute to our discussion of the paranormal? I believe it does so in a significant way. Randy Alcorn summarizes the biblical view…

> This is reality, not myth—we are actually being watched, hovered over, and whispered to, not only by

---

[58] Charles C. Ryrie, *Basic Theology*, (Chicago: Moody, 1999).

God and righteous angels, but also by fallen angels, demons.[59]

Are there angels at work today? Although some would say the Holy Spirit could now fill their role, there is no indication that God has stopped using his divine messengers. The better question might be whether that which appears to be angelic activity is as it seems.

Scripture reveals that righteous angels can and do interact with humans. The New Testament records several angelic appearances.[60] In their true form, they seem to have a fearful effect on observers. (Thus, their typical first words: "Fear not!"). But we're also told that angels might look quite human. Lot's angelic visitors seemed to have human form, and Hebrews 13:2 advises that sometimes we might meet angels and not be able to distinguish them from ordinary people.

But could the opposite also be true? Could a being appear as a righteous angel and not be one? According to Mohammed, an angel claiming to be Gabriel appeared to him in a cave and gave him verses for the Quran. Similarly, an angel named Moroni guided Joseph Smith to artifacts used in the founding of Mormonism. I'm also reminded of the hundreds of tabloid headlines and best-seller books that describe modern-day angelic visits, bringing proclamations of peace and goodness. Talk show guests joyfully spread the "new gospel" they received from an angel. Obviously, these are godly angels—right?

On the contrary, we're soberly cautioned that Satan "masquerades as an angel of light."[61] So an observer might

---

[59] Randy Alcorn, "What Role Can Satan and Demons Have in the Life of a Believer?" www.epm.org, November 5, 2018.

[60] Angelic appearances: Birth of John the Baptist (Luke 1:5-25), wrestles with Jacob (Genesis 32:22-32), with Gideon (Joshua 5:13-6:27), the fiery furnace (Daniel 3:1-30), leads Peter out of prison (Acts 12:1-19), Gabriel speaks to Mary (Luke 1:26-38), tells the disciples Jesus will return (Acts 1:1-11), announces birth of Jesus (Luke 2:8-20), announces Jesus has risen (Matthew 28:1-10).

[61] 2 Corinthians 11:14

perceive a manifestation of Satan as a benevolent, helpful entity. The implication is that demons can do the same. What an ominous, yet important fact! *An entity with a friendly, helpful demeanor could be a deceiver…*

Thus, our impressions of an entity might be totally wrong. If a demon manifests itself, it could (and probably will) appear to be a godly angel. Indeed, who would stick around and listen to a demon who appeared to be evil and scary? Then how can we possibly discern between the true and the deceptive? God's Word provides a tool.

> …do not believe every spirit, but test the spirits to see whether they are from God, for many false prophets have gone out into the world. By this you know the Spirit of God: every spirit that confesses that Jesus Christ has come in the flesh is from God, and every spirit that does not confess Jesus is not from God. This is the spirit of the antichrist, which you heard was coming and now is in the world already (1 John 4:1-3).

This isn't a quick-fix garlic-necklace for warding off evil spirits. "Testing" a spirit might require much time digging deeply into Scripture and parsing biblical principles. It would be even more effective if that preparation was preemptive. Equip yourself for the test in advance. Know God's Word well enough to recognize deception in as soon as it rears its head.

The notion that fallen angels other than Satan might appear as angels of light brings us back to a point mentioned earlier. Satan isn't omnipresent, or everywhere at once, as is God. It's assumed that his widespread work is aided by his fallen minions. If the original angelic multitudes were innumerable, then a third of them would still be a great number. (Perhaps there are enough to assign a few to each person on earth?)

Also, in the New Testament, we often find that demons can *inhabit* some humans. A classic case is found in the Gospel of Luke…

> When Jesus had stepped out on land, there met him a man from the city who had demons. For a long time he had worn no clothes, and he had not lived in a house but among the tombs. When he saw Jesus, he cried out and fell down before him and said with a loud voice, "What have you to do with me, Jesus, Son of the Most High God? I beg you, do not torment me," (Other accounts add the phrase "before our time.")
>
> Jesus then asked for his name.
>
> And he said, "Legion," for many demons had entered him. And they begged him not to command them to depart into the abyss. Now a large herd of pigs was feeding there on the hillside, and they begged him to let them enter these. So he gave them permission. Then the demons came out of the man and entered the pigs, and the herd rushed down the steep bank into the lake and drowned. (Luke 8:27-33).

Often called demon *possession*, this passage provides important insight into the nature and work of demons. They certainly changed the man's behavior, presumably having complete control over his actions. The parallel account in the Gospel of Mark adds that the man could break chains with superhuman strength when people tried to bind him. Also, the demons not only recognized the man, Jesus, they readily admitted his divinity.

Undoubtedly, Jesus wielded the authority to do as he willed with the unclean spirits. This large group of demons (possibly thousands judging by the name *Legion*) knew their ultimate destination. They begged not to be cast into the Abyss at this particular time. The full meaning isn't clear, but Jesus postponed their sentencing and allowed them to enter the pigs.

Jesus cast demons out of many people during his earthly ministry. In most cases, the demons were the source of some physical or mental malady. When Jesus referred to a demon, there is no indication that he was using a metaphor or euphemism. He spoke of *real, spirit personalities*. One wonders why this phenomenon doesn't seem as prevalent today. Are some physical and mental afflictions still caused by demons?[62] The very question conjures Hollywood images of witch burnings and exorcisms. Theologians suggest that demon activity might have been more concentrated during the ministry of Jesus. Or perhaps demon expulsions best illustrated Jesus's authority over Satan.

Our local view of demonic activity could be skewed. Many missionaries report that "possessing demons" are alive and well. The accusation by skeptics is that demon possession is directly proportional to the lack of modern scientific knowledge. But according to Unger, the phenomenon is more correlated with the lack of Christian influence in a region.[63] Demonic activity does appear to be more straightforward and obvious in cultures that would respond to that manifestation. Perhaps demons, seeing disbelief in the biblical supernatural find new, more appropriate ways to engage the "enlightened" population. Unger writes…

> With the crude savage, Satan may best accomplish his purpose as a "roaring lion," inspiring dread and base fear, but with the cultured and educated, in a so-called

---

[62] Clearly much mental illness has organic, physical causes. People with any symptoms of mental illness should always consult a pastor, Christian pschologist, and/or medical professional. Nevertheless, this doesn't preclude the fact that some cases could be like those of Jesus's day.

[63] Merrill F. Unger, *Biblical Demonology: A Study of Spiritual Forces at Work Today*, (Grand Rapids, MI: Kregel, 1994), 82-83.

Christian society, he can often work more effectively when disguised as "an angel of light."[64]

The question often arises as to whether a demon can inhabit a Christian. There is no biblical account of this, although believers can certainly be influenced and tempted. Scripture assures: "We know that everyone who has been born of God does not keep on sinning, but he who was born of God protects him, and the evil one does not touch him."[65] Furthermore, I don't believe the indwelling Holy Spirit would ever have a demon for a "roommate." That being said, the Bible consistently warns Christians to guard against Satan and demonic spirits. Any part of our lives not controlled by God's Spirit is certainly vulnerable to demonic influence.

How Do Satan and His Cohorts Affect Us?

Our objective here, however, isn't so much to debate the extent of demon activity but to note its characteristics revealed in Scripture. I believe the following truths are important in examining other paranormal phenomena…

- The biblical accounts involved *real* demons. There is no hint that Jesus was merely accommodating the mistaken beliefs of the crowds.
- Demons *can* inhabit humans and express their will through the hosts.
- Indwelling demons can influence a person's thoughts, perceptions, behavior, and even their physical abilities.

---

[64] Unger, 82-83.
[65] 1 John 5:18

- Demons respond to authority and can do nothing unless God permits it.

Jesus summarized the Enemy's *modus operandi* quite succinctly…

> He was a murderer from the beginning, and does not stand in the truth, because there is no truth in him. When he lies, he speaks out of his own character, for he is a liar and the father of lies (John 8:44).

Satan's forte, his go-to strategy, his primary skill—LYING. I believe this is possibly the basic axiom in understanding many paranormal phenomena. Although spirits *could* appear to us in some physical sense and *could* possess humans and animals, I think these phenomena are less common than more indirect interactions. As Jesus stated, Satan is adept at manipulating minds with untruth.[66] It's unclear as to whether he can know the exact thoughts of humans. I have my doubts. Nevertheless, after all these centuries, he's certainly an expert on humans. He knows us well. At the least, he has a super-educated guess at what we might be thinking.

We do know that Satan (and demons) can somehow "place" thoughts. Not full-blown mind control, his manipulation seems to be more along the lines of "suggestion." It has been described as a "whisper" or an "impression." Since It's usually so subtle, and it's within a person's own thoughts, Satan's urgings usually don't seem foreign. Here is the shortlist of ways in which Satan does this to people in the New Testament…

[66] When I say that Satan does this or that, I realize that it more likely is a demon that is involved. Satan is not omnipresent, and it's unlikely he spends much of his time on insignificant me. There are plenty in the lower ranks to do his bidding.

- Hindering (1 Thessalonians 2:8).
- Tempting (Matthew 4:10, 1 Corinthians 7:5).
- Keeping from truth (Mark 4:15).
- Causing to say things (Mark 8:33).
- Placing lies in the mind (Acts 5:3).
- Causing sickness (Luke 13:16).
- Performing signs and wonders (2 Thessalonians 2:9).

Does Satan have the ability to play more tangible mind games? Early church leaders believed he could. Augustine wrote…

> What men can do with real colors and substances the demons can very easily do by showing unreal forms...[67]

Luther added…

> Satan then is well able to affect all the senses so that a man would swear he did see, hear, touch, a thing which notwithstanding he doth not…[68]

I would argue that true followers of Christ cannot be *completely* deceived, but there are plenty of nonbelievers around, influenced by fallen spirits and ready to engage believers. Likely, there are many Christians susceptible to Satan's lies by way of deceived leaders, parents, or spouses.

I believe another important principle when dealing with paranormal phenomena is the fact of "virtual reality," not in the techno-digital sense but in the human mind. Sometimes the mere *belief* in a phenomenon alone can be highly effective in the work of the Evil One. Satan's whispered commands may

---

[67] Augustine, *The City of God*, Book 18:5.
[68] Luther, *Commentary on Galatians*, Chapter 3.

need no physical substance: "Fear this ghost… listen to this psychic… ponder this UFO…" How many lives have been affected by simply dwelling on supernatural possibilities?[69]

The question also comes up as to whether Satan can manipulate the material world of atoms and energy. Scripture does indicate the possibility. When Jesus calmed the storm in Mark 4, he "rebuked" the storm verbally. Some take the storm to be an act of Satan. A clearer, "behind-the-scenes" example is in the book of Job. God permitted Satan to test the righteous man, Job. Besides manipulating the minds of two terrorist groups to attack, Satan directly used material means to wreak havoc on the man. He caused fire to fall and burn Job's sheep. Then, he used a mighty wind to collapse the house on all Job's children. In a final intimate and very physical attack, Satan inflicts Job with painful sores. Clearly, Satan can use matter and energy but take note—*only as God permits*.

A related question is whether fallen spirits can clothe themselves in physical bodies as Paul describes our spirit-body duality. In other words, can demons take on a physical body (not simply inhabit a preexisting human body) and interact in our physical realm? We'll address this more fully in our discussion of Genesis 6.

Unpacking the Ephesians Hypothesis

As we examine paranormal phenomena, we'll consider the Ephesians Hypothesis along with any other viable explanations. Investigations of alleged paranormal events shouldn't begin by assuming a supernatural source, but the option must *not* be pushed aside merely because that view might be unpopular in contemporary thinking. This should especially apply to Bible-

[69] This is a danger in any focused study of the paranormal, but the believer has this proven promise "He who is in you is greater than he who is in the world" (1 John 4:4).

believing Christians. Biblical language scholar Michael Heiser writes…

> My contention is that, if our theology really derives from the biblical text, *we must reconsider our selective supernaturalism and recover a biblical theology of the unseen world.* [70]

He goes on to say that this doesn't mean every scriptural narrative involves supernatural elements. But accurate interpretation must recognize that the inspired writers understood and accepted the reality of a supernatural realm.

Scripture plainly portrays Satan as a formidable enemy. He isn't the impish clown of medieval lore, aimlessly vexing humans for his demons' amusement. Unfortunately, this is the strawman the secular world ridicules when Christians discuss the reality of the Enemy. Lucifer was created "the anointed cherub," wise and powerful. He retains some of those traits, thus, his plans are well-ordered and precise. All that he does is intelligent, purposeful, cunning, and targeted to a specific end. Any deception that can advance his plan becomes part of his immense arsenal. I believe we may say, without paranoia or sensationalism, that anything our feeble minds might conceive of as a means of deception, Satan has already exploited it. Paul warns that we should *expect* to experience Satan's attacks…

> Now the Spirit expressly says that in later times some will *depart* from the faith by devoting themselves to deceitful spirits and teachings of demons (1 Timothy 4:1).

An extremely important concept regarding demons and the paranormal is that *anything goes*. Satan isn't bound to any code of ethics in his work (although he certainly is limited by the

---

[70] Heiser, *Unseen*.

permissive will of the Most High). If fear works in deceiving an individual, then fear it will be. If frivolity can lead away from God, then Satan will use frivolity. If one step back brings two steps forward, then he might appear to lose temporarily. If scientific reasoning can carry the lie, then scientific reasoning will be played to the utmost. The diversity (or apparently random chaos) we observe in truly paranormal events might be explainable when the Deceiver has a diverse arsenal and this win-at-all-costs motto.

I suppose that a deceiving spirit would be just as satisfied for their victim to believe in a ghost, ET, monster, or fairy as any other fable that would lead them away from truth and faith. Yet again, we certainly must not make the mistake of seeing the demonic in every incident. We inhabit fallen, fleshly bodies navigating in a sinful world system. There's plenty of evil to go around without blaming everything bad or strange on Satan. This extreme has led to all types of paranoia and fanaticism in the church and has likely served Satan quite well in the ensuing confusion.

But the other extreme is perhaps more dangerous. I can't find any scriptures that advise us to "chill out and stop being so alert!" Instead, we are continually urged to be battle-ready. This isn't passive advice. It's a strategic call to arms against a real and imminent enemy. An accurate, balanced view of the supernatural has always been important, but it's vital today as post-modern misinformation proliferates, even in churches.

Test the Spirits

> Do not believe every spirit, but test the spirits to see whether they are from God, for many false prophets have gone out into the world (1 John 4:1).

A presupposition within this verse is that at least two types of spirits exist and that they interact with this physical world. Some spirits are from God. Some are not. The command to "test the spirits" implies that their natures and motives are discernable, yet not necessarily obvious. There is absolutely no hint in Scripture that even the most mature believers are immune to false teaching and deceptive spirits. Our modern culture demands that we stay open-minded and tolerant toward new ideas. There is a bit of general truth in that, but a balanced stance will maintain a degree of skepticism and discernment anytime a new "message from God" confronts us.

Tozer provides several questions to ask when encountering a "new" message, miracle, prophecy, or doctrine…

1. How does it affect my concept of God and my attitude toward him?
2. How has my attitude toward Jesus been affected?
3. How does it affect my attitude toward God's Word?
4. What is its effect on my self-life?
5. How does it affect my relation to and our attitude toward our fellow Christians?
6. What is its effect on my attitude toward sin?[71]

These guidelines are good, but they assume a working knowledge of the components. This is why we must "vaccinate" ourselves and our families with the solid spiritual disciplines of Bible study, repentance, purity, and prayer. We must proactively possess true concepts about God, Jesus, Scripture, and sin. Unless we have made the effort to "hide God's Word in our heart" we are vulnerable to any spirit that attempts to communicate.

---

[71] A.W. Tozer, *Man: The Dwelling Place of God*, public domain.

7
Signs and Wonders

"Is not the gospel its own sign and wonder? Is not this a miracle of miracles, that 'God so loved the world that He gave His only begotten Son, that whosoever believeth in Him should not perish?' A truthful Saviour ought to be believed. He is truth itself. Why will you ask proof of the veracity of One who cannot lie?" Charles Spurgeon[72]

"We all have Naturalism in our bones and even conversion does not at once work the infection out of our system." C.S. Lewis[73]

How do modern-day miracle claims fit within our previous classification system? Excluding providential events, we'll consider only direct miracles—*events in which a supernatural being blocks at least one natural cause in bringing about a particular situation within a logical context.*

As we examine modern-day miracles, we once again are faced with the extremes of those who shun the supernatural and others who find the miraculous under every stone. This is

[72] C. H. Spurgeon, "Except Ye See Signs and Wonders, Ye Will Not Believe," *Morning and Evening,* September 2nd, Evening.
[73] C.S. Lewis, "Miracles," *The Complete C.S. Lewis Signature Classics*, (New York: Harper Collins, 2002).

especially true in the Church. At one end are those who believe that having to "own up" to the miracles in the Bible is an obstacle for engaging the secular culture. Part of this mindset could also come from the fear of being associated with the other extreme. This group expects, and seeks miracles as a part of worship and life in general. They tend to view the previous group of believers as spiritually cold and unresponsive.

Are Modern Miracles Biblical?

The basis for both extremes relies primarily on their view of New Testament *gifts of the Spirit…*

> For to one is given through the Spirit the utterance of wisdom, and to another the utterance of knowledge according to the same Spirit, to another faith by the same Spirit, to another gifts of healing by the one Spirit, to another the working of miracles, to another prophecy, to another the ability to distinguish between spirits, to another various kinds of tongues, to another the interpretation of tongues (1 Corinthians 12: 8-10).

Both groups agree that in the early church, some were given the spiritual gift of miracles. In the context of the work of the Church, miracles validated the message of the gospel. Although these signs ostensibly benefitted the recipients in some way, in no case was this the precipitating cause. People were better off after Jesus healed them, but the miracle-working of Jesus was primarily for the purpose of establishing his authority and messiahship. When John the Baptist asked whether he was truly the Messiah, Christ responded, "Tell John what you hear and see: the blind receive their sight and the lame walk, lepers are cleansed and the deaf hear, and the dead are raised…"[74]

The connection between miracles and evangelism is explicitly noted many times in the New Testament. When Peter heals Aeneas (Acts 9:34-35), Luke writes, "…the residents of Lydda and Sharon saw him, and they turned to the Lord." Luke also comments on Tabitha being raised from the dead, "It became known to all Joppa, and many believed in the Lord." Referring to these purposeful, directed signs, John MacArthur says…

> A sign is to point to something. [New Testament miracles] were signs authenticating a divinely-commissioned teacher. When God supernaturally—superhumanly—suspended natural law and acted in human history, He did so as a sign, to point to a teacher who was speaking for Him.[75]

The question, however, is whether these *charismata* (spiritual gifts such as miracles, tongues, and healing), common at the time of the newborn Church, are still being bestowed on Christians today. *Cessationists* maintain that signs and miracles ceased when the New Testament scriptures were accessible by local congregations. Warfield explains…

> God the Holy Spirit has made it His subsequent work, not to introduce new and unneeded revelations into the world, but to diffuse this one complete revelation [the Bible] through the world and to bring mankind into the saving knowledge of it.[76]

---

[74] Matthew 11.
[75] John MacArthur, "Does God Do Miracles Today?" *Sermons 90-56*, www.gty.org, August 11, 1991.
[76] B.B. Warfield, *Counterfeit Miracles*, (Edinburgh: The Banner of Truth Trust, 1918).

Many scholars believe the phenomenon of the charismata tapered off gradually, perhaps into the 4th century. Miracles may have been useful in isolated areas and specific situations. Early church leader Augustine wrote...

> For when the [Church] had been diffused and established through the whole world, these miracles were no longer permitted to continue in our time, lest the mind should always seek visible things, and the human race should be chilled by the customariness of the very things whose novelty had inflamed them.[77]

Later, Augustine revised this belief saying that he had witnessed some miracles that were of the New Testament type. However, he noted that these were not comparable to those of the Apostles and Jesus.[78] The point here is that there wasn't the blanket acceptance of miracles in the early centuries that some would have us imagine.

Today's Pentecostals and other charismatic congregations believe that the gifts never completely ceased. Most Pentecostal groups trace their modern origins to evangelical revivals in the U.S. and Britain during the 19th century.[79] Their common belief was that the Church had grown cold over the past two millennia and that God was once again pouring out his Spirit. Today, these groups, anticipating Christ's imminent return, believe that charismata are bestowed when Christians experience the baptism of the Holy Spirit. Worldwide, Pentecostalism is one of the fastest-growing portions of Christianity. The New Apostolic Reformation (NAR) churches are a more recent charismatic movement with a strong emphasis on reestablishing the early church traditions, including supernatural signs and wonders, apostolic governance, and extra-biblical revelation.

---

[77] Augustine, *On the True Religion*.
[78] Augustine, *Retractations*.
[79] The term *Pentecostal* refers to the bestowing of gifts on the early church at the Day of Pentecost gathering.

Some other evangelicals deny that miraculous gifts ceased after the apostolic age.[80] However, many of these do not see these events as standard practice for the church. John Piper argues that modern miracles are possible, but they must conform to Paul's example…

> Miracles were no contradiction of the theology of the cross which Paul proclaimed and practiced, since they were performed not in a context of triumphant success and prosperity, but in the midst of the distress and vilification he was obliged to endure.[81]

MacArthur has a more limited view of contemporary miracles as a standard part of Church experience…

> His power has not diminished in the least since the days of the early church… if God chooses to do something miraculous, He can do it. I am convinced that most of the miracle signs and wonders—if not all—being claimed today in the charismatic movement, have nothing in common with what we know about biblical miracles. They don't fit the biblical criteria.[82]

---

[80] C. Peter Wagner and John Piper are mainstream evangelicals who believe the charismata have not ceased.
[81] C. K. Barrett, *Commentary on the Second Epistle to the Corinthians*, (New York: Harper and Row, 1973), 321.
[82] MacArthur, "Does God Do Miracles Today?"

Miracles and Catholics

The Catholic Church (CC) has historically experienced its own brand of miracles.[83] Not only are supernatural events a significant factor for a large number (one billion Catholics worldwide), the phenomena have become part of the culture in general. Many non-Catholics are aware of some of their miracles and have incorporated them into their worldviews.

As a Protestant (and Evangelical) I admittedly have some views that vary significantly from the teachings of the CC. My intention here isn't to debate or demean specific doctrines or members, but to address the nature of miracles associated with Catholics. I'll depend primarily on sources within their documents.

The CC's understanding of the supernatural is in many ways more clearly defined than that of many evangelical groups. As a doctrine, they *expect* that supernatural signs will be used by the Holy Spirit in the development and sustenance of faith…

> That the submission of our faith might nevertheless be in accordance with reason, God willed that external proofs of his Revelation should be joined to the internal helps of the Holy Spirit.[84]

Indeed, denying the possibility of supernatural intervention by God would place one out of communion with the Church…

> If anyone shall say that miracles are impossible, and therefore that all accounts regarding them, even those contained in Holy Scripture, are to be dismissed as

---

[83] I use the term Catholic Church (CC) to refer to most of the rites in general. The examples and quotations are primarily from the Roman Catholic Church, since it's likely recognized by more readers.

[84] Catechism of the Catholic Church, *Dei Filius* 3

> fables or myths; or that miracles can never be known with certainty, and that divine origin of Christianity cannot be proved by them; let him be anathema.[85]

Thus, a practicing Catholic would anticipate that miracles would be common among their billion members. They are quick to note that a true miracle already occurs with each individual communion event…

> Miracles big and small surround us, including the greatest one that happens every hour of the day in every country of the world: the Eucharist, bread and wine transformed into Christ's body and blood…[86]

The CC emphasizes that supernatural interventions are meant to be signs or revelations pointing to a truth greater than the outcome of the particular miracle. It also has a logical way of classifying miracles at the base level. The system shows a clear differentiation between biblical miracles and those performed later…

> The miracles of Christ and the subsequent works of the apostles in his name come down to us through Sacred Scripture, which is considered to be *public* revelation, as it's valid for all time and meant for all. Miracles and messages received after the death of the last evangelist, John—even extensively studied and Church-authorized spiritual insights given to history's greatest saints—are considered *private* revelation.[87]

---

[85] Vatican I, Session 3: *Dogmatic Constitution on the Catholic Faith*, www.papalencyclicals.net, Accessed Feb 24, 2020.
[86] *Lumen Gentium*, Dogmatic Constitution of the Church, no. 5.
[87] Michael O'Neill, *Exploring the Miraculous,* (Huntington, IN: Our Sunday Visitor, Inc., 2015).

Thus, all the miracles recorded in the Bible are *public revelation*, as are all other truths found in Scripture. Catholics, as well as conservative Protestants, would agree on this aspect. The questions, however, concerns *private revelation.* As previously discussed in an evangelical context, are these signs needed in post-apostolic times? Pope Benedict XVI believes they are…

> Private revelation is an aid to [faith in the Scriptures], and it demonstrates its credibility precisely because it refers back to the one public revelation.[88]

As for relevance to present-day believers he goes on to say that personal revelation can…

> …introduce new emphases, give rise to new forms of piety, or deepen old ones. It can have a certain prophetic character and can be a valuable aid for better understanding and living the Gospel at a certain time…

Despite the popular view that the CC actively exaggerates molehill events into mountains, Catholic leadership is well aware that a miraculous claim is very likely not a true miracle. In fact, church leaders wisely note that presumed miracles may become a distraction to religious practice…

> Excessive, obsessive expression of belief by the faithful in miraculous phenomena is not only the reason the Church is methodical and cautious in approving any occurrence as authentic but also a primary impetus for performing any investigation in

---

[88] Joseph Cardinal Ratzinger, Post-synodic Apostolic Exhortation *Verbum Domini*, no.14.

the first place. The unspoken goal of such examination is to prove that nothing supernatural is occurring at these places, in order that the faithful might return to a more authentic and grounded practice of their faith.[89]

Due to the flood of miracle claims in the 1900s, the Vatican created a document laying out guidelines for discerning private revelations. In 1978, Pope Paul VI gave local bishops the responsibility of investigating the events. It also provided guidelines for discerning supernatural intervention as summarized by O'Neill…

- The facts in the case are free of error.
- The person(s) receiving the messages is/are psychologically balanced, honest, moral, sincere, and respectful of Church authority.
- Doctrinal errors are not attributed to God, the Virgin Mary, or a saint.
- Theological and spiritual doctrines presented are free of error.
- Moneymaking isn't a motive involved in the events.
- Healthy religious devotion and spiritual fruits result, with no evidence of collective hysteria.[90]

Should a private revelation fail any of these criteria, Catholics are required to recognize the decision of the ecclesial authority and not regard the event/object as authentic. Interestingly, a church authority may reject the actual

[89] Michael O'Neill, *Exploring the Miraculous,* (Huntington, IN: Our Sunday Visitor, Inc., 2015).
[90] O'Neill, *Exploring.*

revelation, yet still, sanction the public veneration surrounding the event. In other words, the miracle may be false, but any good fruits resulting from the event (increased faith, repentance) are considered acceptable. If a private revelation is discerned to be genuine, it's deemed *constat de supernaturalitate*.[91] At this point, church members may believe the revelation, however, this doesn't mandate an obligation to do so.

Currently, healing miracles are also being scrutinized more closely within the CC. For example, the alleged healings at the holy site in Lourdes, France are reviewed by a medical committee before accepted as authentic. Dr. Franco Balzaretti explains that in order for a miraculous cure to be authenticated, it must be 1) clear, 2) instant, 3) complete, 4) permanent, and 5) demonstrated with medical documentation before and after. The healed condition also had to have been serious and unable to be cured by normal means.

One of the most common Catholic miracle is a *Marion apparition*. Catholic theology claims that Christ's mother, Mary, was taken bodily into heaven, thus she is able to manifest herself spiritually (or bodily) in order to aid the Church. There are many church-authenticated visitations by Mary. A *visionary* may literally have a vision of the Virgin, hear a spoken voice, see a seemingly physical apparition, or visualize an image of her in an object.

Perhaps the most well-known Marian apparition occurred in Fatima, Portugal in the summer of 1917. The apparition known as *Our Lady of Fatima* appeared to three young children on six different occasions, on the 13th of each month. The children claim to have been told many things by Mary and an angel, including her request to "pray a lot, a lot for the sinners and sacrifice a lot, as many souls perish in hell because nobody is praying or making sacrifices for them."[92]

---

[91] Latin for "It is established that this is something supernatural."

[92] John De Marchi, *The Immaculate Heart,* (New York: Farrar, Straus and Young, 1952).

The apparition informed the children of a final miracle to come on the next October 13th. It was reported that up to 100,000 people may have gathered at the site that day. In what is also called the *Miracle of the Sun* many claimed to have seen the sun radiate colorful beams while shooting back and forth across the sky. Others say they saw nothing unusual. In 1930, the local Bishop declared the Fatima events "worthy of belief." John Paul II had great respect for Our Lady of Fatima, regarding an incident in his life…

> At that time I experienced mortal danger and suffering, but also the great mercy of God. By the intercession of Our Lady of Fatima my life was given back to me.[93]

Other manifestations of Mary haven't been as well-received by the CC. Leaders bemoan having to deal with the pareidolia of those seeing holy images in everything from clouds to potato chips.[94] Often artworks of Mary or one of the saints are reported to have blood or tears oozing from the image…

> Of these images of our Holy Mother, some have been identified as being miraculous. It is not that the statue or painting is miraculous of itself, but it does seem that Our Lady favors certain of her replicas and often honors the requests of those persons who visit them to express their needs and their love for her.[95]

---

[93] Consecration of the Church of Our Lady Fatima, Homily of John Paul II, *June 7, 1997.*
[94] A boy in Miami found the image of Mary on a grilled-cheese sandwich. His mother sold it on Ebay for $28,000.
[95] Joan Carroll Cruz, "Miraculous Icons," Excerpted introduction to *Miraculous Images of Our Lady,* www.the miraclehunter.com, Accessed Feb 2, 2020.

The skeptical side of me says that people are usually bad at discerning miracles. Certainly, God can and does work miracles whenever and however he chooses. Yet I'm guessing a large proportion of modern miracles fall into the *mistaken identification* category, the majority of these being merely statistical phenomena. *Events of low probability may simply happen, given enough time and a large pool of events.* I've personally witnessed charismatic events in which participants are very likely experiencing something more akin to psychology than to phenomenology.

We might also expect some miracles to be hoaxes. In churches alone (Catholic and Protestant), we have witnessed scandals engineered for the profit or prestige of the parties involved. These we'll classify as *directed human activity*. This doesn't necessarily have to involve deliberate trickery. An overworked sitcom example… Suppose you witness a murder, then later see the victim walking around in good health. Oh, what you didn't know was that you had observed a group rehearsing a scene from a play. From your perspective, though, it seemed a "miracle" had occurred.

At this point, I probably seem quite cynical. I've bashed miracles and accused some of my brothers and sisters of deception. I would argue that Scripture commands us to be discerning. Miracles are rare. As one wise person said, "That's why they're not called *normals*."

But now perhaps we can more effectively examine that small group that remains—those miracles that occur with very low probability *and* that exist within a context of real narratives and events. Instead of starting with modern miracles, let's see how truly *bona fide* miracles—those performed by Jesus—display unique characteristics.

In chapter 9 of John's gospel, Jesus heals a blind man. We are told this man has been blind since birth which implies that his condition was long-term, incurable by conventional treatment, and well-known by the community. Even today with modern technology, there is little chance that total blindness can be cured, so his restored sight was an extremely low probability. There might be mathematicians who still claim that the event is improbable yet possible by normal means. But there is more involved in this miracle.

As with the baby bag example, the context of this miracle is more convincing than the mathematics. Consider the big picture... A rabbi claiming to be God happens to come across the blind man. The local group of religious leaders needed to learn a spiritual lesson. Twelve disciples needed training in the supernatural. A community (who had spent years dealing with the blind man and knew him well) needed faith in a Savior. And possibly most significant, God had decided a couple of decades earlier that a baby would receive a defect so that all of this could come together at a particular place and time.

In another well-known miracle, Jesus feeds thousands with a basket of fish and bread. Skeptics are quick to dismiss this account simply because it's *too* miraculous, so it's a good example to cite. I believe it would require an even greater miracle to fake or cause misidentification of this event. The probability of fooling one person might be calculable, but I would assume close-up sleight-of-hand involving twelve servers and multiple thousands pushes the limits of probability.

Once again, several components within a context take even a low probability to the next level... A situation occurs in which thousands of people have a physical need. Twelve disciples need more hands-on instruction. A crowd of sinners at a particular time and place need faith in a Savior. God desires an example that will inform for centuries and glorify him for eternity. Now that's a miracle!

Discerning a True Miracle

Are modern miracles fake or just extremely rare? I believe I can say with certainty that true miracles are at least rarer than most people imagine. But I would never want to be in the position of boxing God into a quota, based on my flawed human reasoning. I believe the "safest" approach would be to hold the miracle in question up to the light of genuine New Testament miracles. Here are some characteristics of the miracles recorded in Scripture. This isn't meant to be a checklist, but rather a qualitative model that reveals by its uniqueness the counterfeit.

- Biblical miracles were easily and naturally attributable to God. Although God may have used people as channels, there was never a question as to the source.
- Obviously, the recipients of miracles benefitted, but biblical miracles were intended *primarily* for the benefit of the observers. Furthermore, the ultimate benefit was the spread of the gospel.
- Other than enhancing their authority to spread the gospel of Christ, miracles did not benefit the one performing them.
- Miracles did not contradict Scripture in any way, nor did they add information to Scripture.
- The results of a miracle were immediate and permanent (for the life of the recipient).

Compared with biblical accounts, many modern "miracles" fall miserably short on one or more of these points. Believers must guard against "Godless myths and old wives' tales" that ultimately lead away from God rather than bring him glory. God can and will use miracles as he pleases, but he doesn't need the assistance of even our well-meaning stage shows.

8
Ghosts and Other Things that Go Bump in the Night

"There's more of gravy than of grave about you, whatever you are!" Ebenezer Scrooge

"...People are destined to die once, and after that to face judgment." Hebrews 9:27

Poltergeist, phantom, shade, wraith, spook, bhut, yurei, specter, banshee. The names reflect the diversity of cultures that have entertained the idea that some essence of life continues after death. The belief is rooted somewhere in the shadows of our prehistory. From Gilgamesh's encounter with his dead friend, Enkidu, to A&E's *Ghost Hunter* series, belief in the spirit world intrudes into our world endemically. According to a 2009 survey of U.S. adults, 18% claim to have personally experienced a ghost. About a third say they have felt in contact with someone who has died.[96]

The media abounds with entertaining depictions of ghosts and the afterlife. We get fascinated—they get rich. According to the general storyline, at death there is a lighted tunnel through

[96] Pew Research Center for the People & the Press and the Pew Forum on Religion & Public Life, 2009.

which a spirit is compelled to pass, otherwise, a popular modern psychic explains…

> A ghost is created when, at death, the spirit either sees the tunnel and turns away or refuses to acknowledge the tunnel in the first place, with the result that it gets caught, outside of its body, between our dimension and the dimension of The Other Side.[97]

This ghost recipe is essentially a throwback to some very ancient ideas. By the Middle Ages, some in the church had compromised a hefty portion of the pagan folklore of earthly spirits. This doctrine of purgatory did much to shape our western ghostly traditions. As the doctrine was refined over the next few centuries, a "Christianized" explanation for those restless spirits was popular...

> Because if they die truly repentant in charity before they have made satisfaction by worthy fruits of penance for sins committed and omitted, their souls are cleansed after death by purgatorical or purifying punishments… [98]

Since few were likely to achieve these "worthy fruits," multitudes of suffering souls could now be postulated, moving in and out of the earthly realm, striving to purge their sins and gain entrance into heaven. Soon, any person dying with unfinished business was believed to be doomed to haunt their former location, obsessed with some desire for "closure." Proponents of this view attempt to provide a theory of how all of this works. But, when we extract the general doctrine espoused, it seems to be a fuzzy, pop-culture pantheism in

---

[97] Sylvia Browne, *Life on the Other Side*, Penguin Books, Ltd. 2000
[98] Second Council of Lyons, 1274.

which the deceased person (soul) never quite has a clear idea of what's happening.

What Does a Ghost Look Like?

The title may seem to beg the question, yet it seems that many people are perceiving something. In his early book on the paranormal, G.N.M. Tyrrell explains…

> "Do you believe in ghosts?" is one of the most ambiguous [questions] which could be asked… "Do you believe that people sometimes experience apparitions?" the answer is that they certainly do.[99]

The term "apparition" is perhaps more appropriate here since it speaks of an *appearance* or *event*. "Ghost," however, implies a personality. To reiterate part of the rationale for this book, what a person experiences is often more significant than what happens in reality. This is in no way a bow to relativism, but rather the acknowledgment that our perceptions and beliefs about an event are what most affect our worldview.

I'll suppress the strong urge to tell a string of ghost stories. There are plenty of Internet sources for anyone reminiscing campfires and s'mores right about now. Instead, let's examine the commonalities among ghostly phenomena.

Tyrrell did his research on the paranormal during the first half of the 20th century. With degrees in math and physics, his studies do seem to be well-documented and systematic. He also claims to have included only "reputable" witnesses in his studies. From his large case file, Tyrrell created what he called the "perfect apparition." This was a theoretical "collage" of the cases he had studied. These features are common, but not

---

[99] G.N.M. Tyrrell, *Apparitions,* (New York: Collier, 1953, e-book 2018).

necessarily present, in each incident. For Tyrrell, the perfect apparition, "while standing beside a normal human being," would exhibit the following …

- Both would appear equally real and solid.
- The apparition could be viewed from any distance or angle chosen.
- The apparition might speak to us or respond, but not at length.
- If a mirror were present, both figures would be reflected.
- Both figures would cast shadows.
- If the apparition was wearing a rose, we might smell it.
- We would hear sounds associated with a real person (shuffling, stomping).
- We might feel its touch, and it would turn its head to follow our movement.
- In addition to clothing, other accessories might be present, such as a cane, watch, suitcase, or even a pet.
- The apparition might manipulate real objects.
- If we tried to hold on to the apparition, we might feel it or our hands might pass through it. We might feel a "coldness" near the figure.
- In time (seconds to a half-hour), the apparition would fade, vanish, or pass through a wall.

In my research of ghostly reports, Tyrrell's insight also seems to coincide with the majority of modern perceptions of the phenomenon. We'll use these claims in future comparisons.

Bible Ghosts

Those who pursue ghostly experiences often point to the book of 1st Samuel as biblical evidence.[100] Amid battle with the Philistines, King Saul isn't getting an immediate response from God, so he decides to take matters into his own hands. Saul goes to the neighborhood medium and convinces her to channel the ghost of the departed prophet Samuel. The Endor medium does her act but seems terrified when the "ghost" of Samuel actually appears. The spirit's first words are "Why have you disturbed me by bringing me up?"

Theologians differ over the interpretation of this incident. Some claim that the spirit was merely a demon impersonator of Samuel. The spirit gives a stern rebuke to Saul which does sound as though he is still prophesying for God. Another view is that in a rare act, God allowed Samuel's spirit to be brought from Abraham's bosom[101] for a special pronouncement to the ungodly king. In either case, Samuel wasn't hovering around waiting to be summoned. This appears to be a unique situation and in no way a justification for the concept of ghosts, channeling, or necromancy. Scripture expressly forbids any attempts to contact the dead…

> There shall not be found among you anyone who burns his son or his daughter as an offering, anyone who practices divination or tells fortunes or interprets omens, or a sorcerer or a charmer or a medium or a necromancer or one who inquires of the dead, for whoever does these things is an abomination to the LORD (Deuteronomy 18:10-12).

---

[100] The word for "spirit" in the Bible denotes "wind" or "breath". Depending on the translation, it may be rendered "ghost" or "spirit". The KJV uses the term "Holy Ghost" for the third person of the trinity.

[101] Abraham's bosom is a term used to designate the place where the spirits of Old Testament saints resided temporarily.

The New Testament references ghostly spirits, but only in the context of colloquial beliefs. When the disciples saw Jesus walking on the water, their mistaken assumption was that it was a ghost.[102] Also, when Peter was miraculously released from prison, those hiding in Mary's house mistakenly believed at first that his appearance was his spirit.[103]

However, nowhere does Scripture indicate that we remain on earth or connected to this physical realm after death. The consensus of God's Word is that at death we instantaneously enter eternity either with Christ in heaven or separated from him in hell. Taken literally, Jesus's account of the rich man and Lazarus implies that returning to the physical world isn't just rare, it isn't allowed.

Our Dual Nature

Despite the multitude of erroneous beliefs, most of us with a Christian worldview readily accept the idea that there truly is a consciousness that survives death. The Bible is quite clear about the destination of human souls. Indeed, that is the hope of Christianity! Our point of contention concerns the *nature* of the afterlife. In one of the clearest descriptions of our supernatural composition, present and future, Paul settles some of our deepest concerns...

> For we know that if the tent that is our earthly home is destroyed, we have a building from God, a house not made with hands, eternal in the heavens. For in this tent we groan, longing to put on our heavenly dwelling, if indeed by putting it on we may not be found naked. For while we are still in this tent, we

---

[102] Matthew 14:26
[103] Acts 12:15

> groan, being burdened—not that we would be unclothed, but that we would be further clothed, so that what is mortal may be swallowed up by life… We know that while we are at home in the body we are away from the Lord… and we would rather be away from the body and at home with the Lord (2 Corinthians 5:1-8).

Paul says that I inhabit my body just as I might dwell in a tent. I am not simply a physical machine as Darwinian evolution would predict. Only my "tent" is earthly. Whatever electrochemical mechanisms are at work in my brain, and however they interface with my real, conscious self, that "me" isn't totally equated with my physical body. In other scriptures, that nonmaterial component is identified as a soul or spirit. *I don't have a soul, rather I am a soul who lives in a body.*

Then Paul mixes in an even more intimate metaphor. I am *clothed* with a body. And, it's conceivable that I could remove the "clothing" and still be a conscious entity. He quickly adds, though, that that would not be the ideal situation. Although God and his angels may exist as pure spirit, he has made my spirit to best reside in a body.

If it concerns you that your physical body might not have a very long shelf life, not to worry! In the future, our painful, deteriorating "clothing" will one day be replaced with new apparel. Although we can't fully understand the details, our new suits will be of an immortal type, fit for eternity.

Ghosts in the Machine

Secular psychology attempts to link the belief in ghosts and the afterlife to our psychological denial that this current life must end. I have to admit that thoughts of my death aren't entirely pleasant, even though my faith anticipates a heavenly home. A wise Christian mentor once told me: "I don't fear death, just the painful stuff immediately prior to it."

But like most of my Christian brothers and sisters, I think my belief in the afterlife goes beyond an "animal" fear of finality. We all have a common aversion to nonexistence, in part due to that inner sense of eternity mentioned earlier. Our souls seem to know intuitively that our true selves are even now transcending physical time and space, though our bodies are connected to this physical realm.

As expected, when serious investigators attempt to organize the data on ghostly events, the results are often so skewed and dissimilar that scientific conclusions are difficult. Perhaps this is a good opportunity to apply the Pickle Principle…

Experience and intuition tell us that many of the phenomena are likely *mistaken identifications*. It doesn't require occult rituals or exotic equipment to see a ghost. As Ebenezer Scrooge observed, his specter was likely "an undigested bit of beef… a fragment of an underdone potato." Would anyone doubt that many ghostly sensations are reported by those who might be under the influence of alcohol, insomnia, a midnight movie, or some other altered mental state? I won't bore you with the list of self-inflicted, Saturday-night apparitions claimed by a guy in my college dorm, but it happens.

It's also possible, if not likely, that another large portion of hauntings is the result of *directed human activity*—or in this case hoaxes. Who doesn't enjoy a good ghost story? And if it brings in tourists and book deals, all the better. Hollywood has shown us that ghostly deceptions aren't only easy, but they are also entertaining. I haven't watched years of *Scooby Doo* for nothing.

The Pickle Principle tells us that even if these explanations are found to dominate, it doesn't necessarily follow that all the occurrences are of this type. This is a form of the *composition fallacy*, in which the whole is assumed to be the same as a smaller portion. Obviously, that isn't always the case. And, an *appeal to ignorance* argument is equally unproductive. The statement *"No one has ever proven that ghosts exist, so they must not be real"* is just as invalid as *"No one has ever proven that ghosts do not exist, so they must be real."*

Scientists also know that it's easier to falsify a hypothesis than to verify it. For example, how would we verify the statement "There are no ghosts?" If we searched every place on earth, a skeptic could still claim "You looked at the wrong time" or "You didn't look on Mars." In any practical sense, the statement cannot be verified. But if one, single genuine ghost could be observed, then that statement is effectively falsified.

We also might consider whether our definition of *paranormal* might fit here. Could apparitions be some strange physical phenomenon that is yet to be explored? The army of modern-day ghostbusters who attempt to detect paranormal activity through "ectoplasmic" gadgets and electromagnetic fields would certainly think so. Some claim ghosts are the result of some yet-to-be-identified "life energy." Proponents often suggest that with more research, the physical nature of spiritual phenomena might be determined.

But let's explore another alternative. Given that the vast majority of ghostly encounters are likely mistakes and hoaxes, is it possible that some of the remaining incidents *could* be explained supernaturally (strictly as *biblical supernatural)*?

Are Ghostly Phenomena Consistent with the Ephesians Hypothesis?

Could the sensory phenomena associated with spirit encounters fall under the purview of demons? Scripture warns that fallen spirits are able to lead thoughts and alter perceptions.

Sights, sounds, and smells are no challenge for an entity able to deceive human senses. Add to this the tendency of humans to embellish and imagine, I wouldn't be surprised if "ghostly haunting" is just an entry-level job for aspiring demons.

It's also important to remember the context of any paranormal phenomenon we witness. The Ephesians Hypothesis operates within a narrative. Consider the general "theme" of ghostly encounters. Does this phenomenon *usually* encourage people? Do ghostly messages help spread the gospel? Do the spirits give praise to the God of the Bible? Do you feel more sanctified or less when you think of ghosts?

On the contrary, when we look at the unexplained incidents of ghostly encounters, the general mood seems to be fear, foreboding, and malice. There might be a few "innocent" encounters, but even with those, the effect is usually to lead away from the biblical worldview. Also, the messages channeled through spirits seem to have a common thread. Here are some general implications that emerge from reported ghostly communications…

- Ghosts are dead humans.
- Several possible paths might be taken after death.
- Ghosts are not very happy. They are confused or striving.
- The afterlife is a vague, irrational realm.
- Reality is New Age, mystic, occult, or pantheistic.
- Biblical views of the afterlife are wrong.

Undoubtedly, ghosts fail to support the Biblical view of the supernatural. We might ask whether demons have anything to gain by deceiving the living. Is there a good reason to impersonate the dead? One possible answer…

> As servants of Satan and enemies of God, they would have every reason to cast doubt on God's

> Word and its warnings about future judgment. Getting people to believe in ghosts automatically forces them to reject key Bible verses. [104]

Furthermore, ghosts seem to be a diverse species. They might appear to be purposeful, confused, happy, impish, miserable, erratic, angry, scared, scary, friendly, innocent, funny, or violent. If they are supposed to be those once living, how does their apparent afterlife experience line up with the biblical view? If Satan's purpose is to oppose God, diminish faith in his Word, and deceive believers, what better way is there? Ghosts defy all that Scripture says about Heaven and Hell.

Let's back up a step. If demons *would* present themselves as ghosts, *could* they? Does the Ephesians Hypothesis have explanatory power with regard to what people experience? Incidents involving alleged ghosts are not dissimilar to the Scriptural portrayal of the supernatural. Ron Rhodes notes…

> People sometimes genuinely encounter a spirit entity—though not a dead human. Some people encounter demonic spirits who may mimic dead people in order to deceive the living. Many who claim to have encountered such spirit entities have some prior involvement in the occult.[105]

Granted, demonic spirits can influence human thoughts and perceptions to some extent. But what about *poltergeists* (literally, "noisy ghosts")? There seems to be a categorical difference between mental deception and actual interaction with the natural world. As discussed in an earlier chapter, there is scriptural evidence that demons can at times manipulate

---

[104] Rick Barry, "Do You Believe in Ghosts?" July 1, 2010. wwww.answersingenesis.org

[105] Ron Rhodes, *The Truth Behind Ghosts, Mediums, and Psychic Phenomena*, (Eugene, OR: Harvest House, 2006).

physical objects. Could poltergeist activity be correlated with demon activity? Apologist John Ankerberg believes it's possible…

> …the poltergeist phenomenon and its occult connection offer strong empirical evidence for the demonic nature of these spirits. In fact, we know of no poltergeist case that cannot be accounted for on the basis of this theory. We believe the demonic theory is rejected today simply because mediums, parapsychologists and others don't like it. [106]

And obviously, physical evidence is more convincing than mere sensory phenomena. "Noisy ghosts" might be more effective deceivers than the ephemeral ones as the evidence isn't simply personal experience.

Given these observations about ghostly encounters, let's compare some common characteristics with scriptural revelations about the spirit world in the Ephesians Hypothesis…

1. Those who witness apparitions experience sensory-rich phenomena that seem to have physical reality. Demons are able to affect human perceptions, and to some extent manipulate physical objects.

2. Apparitions vary in appearance and behavior. Satan works in ways most effective with the individuals being deceived.

3. Ghosts seem to circumvent physical laws. They appear, disappear, and pass through solid objects. Demons reside in another realm but can interact with ours. Their non-physical manifestations would not be subject to physical laws. Nor

---

[106] Dr. John Ankerberg and Dr. John Weldon, *Ghosts/Part 1*, 2005, www.jashow.org

would they necessarily be detected by cameras or other devices, unless they wish.

4. Ghostly phenomena tend to lead away from Scripture and deny God's truth and the biblical worldview. The objective of fallen spirits, as given in Scripture, is to oppose God.

5. Ghosts seem to possess knowledge about the life of the alleged deceased (from which they're supposedly derived) and about loved ones, associates, etc. Neither Satan nor demons are omniscient, but they obviously have observed earthly events (possibly in greater detail than humans) and have insight into our affairs.

6. Apparitions are flirtatious yet elusive. It's in Satan's best interest for his deceptions to be alluring yet mysterious and generally untestable.

The take-home lesson from ghosts is that *Satan will do whatever it takes* to oppose God's Word and deceive his children. He is serious about his goals regardless of our interpretations. Ghostly phenomena (and many other paranormal phenomena) allow deception to be customized. There is no one-size-fits-all in the Deceiver's repertoire. An informed biblical worldview allows us to discern his schemes in whatever disguise they appear. We'll examine ghostly phenomena in greater depth later.

Goblins, Gremlins, and Ghouls

How could we discuss the paranormal beings without considering those childhood nemeses from under the bed? We have seen that fallen spirits may appear as "angels of light" and will use any means within their power to deceive, destroy, and lead astray. The relevant question here is: Would they have anything to gain by presenting evil, threatening, and malevolent images? Childhood imaginations are creative enough, and I pray God gives excess grace to children by protecting them from actual demonic encounters. So, the specific adult question is: *Would demons have anything to gain by presenting themselves as demons?*

We'll examine spirits as objects of worship later. For now, let's limit our view to spirits as objects of fear. If any good aspect of *daemonophobia* (fear of demons) can be found, it might be that at least the victim's worldview acknowledges the supernatural realm. Demonic encounters are very similar to ghostly ones (after all, they have a common source), except in the latter, spirits may pretend to be *other than* their true malicious selves.

"Evil spirits" are another indigenous belief in cultures worldwide. The Christian worldview doesn't deny these entities in principle. Like flood and creation events, it's likely that cultural memory has transmitted these foundation truths from the Eden. Children may worry about goblins and boogeymen, but adult history records a continuing obsession with jinn, wendigo, daimonia, Oni, ifriti, and Nun'Yunu'Wi. As we'll see later, it may be difficult to differentiate perceptions as biological or paranormal.

In areas where animism, shamanism, and folklore already have a strong grip on people's imaginations, evil spirits seem to be more likely to manifest themselves. Psychologists readily dismiss these accounts, since the experiencers already have that component in their belief system. I believe it could be interpreted differently. Could demonic manifestations be more common in shamanistic cultures because they are most effective there? The same phenomenon in the backyard of a naturalistic skeptic would be counterproductive for a demon. It would actually provide the observer with evidence for the supernatural.

Our unfamiliarity with demonic activity in post-modern America isn't a good indicator. Christian missionaries provide ample verification that demonic phenomena are common where paganism is the tradition…

> It's one of those bullet points under the spiritual warfare topic that seesaws between a conversation killer and an otherworldly fascination that seems too mystical to engage… Although demon possession is a topic the US church usually skirts, it's one that's readily engaged by other cultures. [107]

The good news in the midst of this dark reality is indeed *the* "good news." The most hideous and pervasive demon activity is no match for the truth in the gospel message. Demons can work only where there is ignorance of Christ's work.

---

[107] Caroline Anderson, "Demon Possession, Missions, and the Power of Jesus's Name," www.imb.org, 13 Oct 2017.

9
The Occult

"The most dangerous spiritual guide is the person who is 95 percent true to the Scriptures. Remember, it is not the truth that hurts you: rather, it is the evil." A.W. Tozer [108]

"I'm telling you, the Devil is very smart. He's not going to come out with the horns and the tail, he's going to come in as the slickest car salesman you ever saw." Alice Cooper [109]

"Occult" literally means "hidden." The implication is that certain knowledge is hidden from humans and that we would be better off knowing it. More specifically, "God has information and we want it now!" We've already looked at the seed of this idea when Adam and Eve decided to access forbidden knowledge.

Occultism may include several phenomena—fortunetelling, necromancy, magic, or spiritualism. It has close ties to several other ideologies, such as New Age and certain world religions. The occult is a much more complicated system than the image most of us have of a mysterious Gypsy dealing

---

[108] A.W. Tozer, *The Crucified Life*, (Minneapolis: Bethany House, compiled 2011), 189.

[109] Michael W. Chapman, "Alice Cooper, Christian: In Many Songs I'm 'Always Warning About Satan'," cnsnews.com, 29 Sep 2015.

tarot cards in her caravan or a TV psychic. The huge proportion of fakers and con artists in the craft lead many to dismiss the entire subject. However, the biblical worldview must encompass the big picture…

> Not all occultism is genuinely satanic. Human greed and self-seeking account for much of it… Yet Satan stands ready to pounce upon his victims who exhibit interest in the occult, for it is his domain.[110]

In defense of some psychics, I don't believe they are all willfully deceptive. Hucksters are always drawn to fame and easy money, and the psychic career field is overflowing with opportunity. However, we must not be led to believe that all psychic phenomena are human trickery. The Pickle Principle would suggest that some psychic phenomena could genuinely be supernatural. Unger warns…

> Although much that passes for present-day spiritualistic manifestation is pure chicanery, nevertheless, that real communications from the spirit-world are at times received, cannot for one moment be doubted, if both Scriptural and non-Scriptural evidence is to be given credence.[111]

I hope the reader is beginning to notice the principle that things appearing to be random, chaotic, and diverse aren't always indicative of a lack of strategy.

---

[110] C. Fred Dickason, *Angels Elect and Evil*, (Chicago: Moody Press, 1995), 221.
[111] Merrill F. Unger, *Biblical Demonology: A Study of Spiritual Forces at Work Today*, (Grand Rapids, MI: Kregel, 1994), 157.

Séances and Psychics

If you're having a feeling of *déjà vu,* relax. Paranormal phenomena tend to overlap at many points. But considering their source, should we be surprised?

Séances involve conjuring a spirit (presumably of a dead person) for obtaining supernatural information. Thus, we've just connected ghosts, divination, psychic channeling, and necromancy.[112] Typically, the medium (a human channel through which the spirit is contacted) manipulates some type of life energy (participants joining hands for the energy flow) which psychically summons the entity. The spirit either speaks through the medium or performs some poltergeist gymnastics.

The easy part of analyzing this phenomenon is that it's likely 90% fraudulent. I would suspect there are plenty of demons who would like to jump in with some really deceptive messages, but most mediums have enough illusions and stage tricks to carry the sessions without supernatural assistance.

Unfortunately, this type of trickery can lead people to believe that all supernatural phenomena are fake. If genuine supernatural incidents do occur during a séance such as automatic speaking and writing, clairvoyance, and poltergeist activity, these are well within demonic purview. And realistically, Satan can use fake séances as well as supernatural ones in leading participants into error.

As for the other pursuits of career psychics, my intuition says even fewer are genuinely supernatural. (Again, the Evil One can often capitalize on an individual's *belief* in a phenomenon as well as its actuality.) Human minds are easily manipulated, especially when vulnerable and convinced there is a psychic solution to a problem. As Obi-Wan so wisely noted, "The Force can have a strong influence on the weak-minded." Successful psychic con artists are experts at what they do. A skilled "psychic entertainer" can play technology, old-

---

[112] A dangerous combination, as King Saul might attest.

fashioned manipulation, and cold reads like Paganini on a Stradivarius. Bottom line—whether a psychic phenomenon is genuine or fake, it isn't an option for the obedient follower of Christ.

Spiritualism

Satan seems to work diligently where God is working greatly. Many had come to faith during the Second Great Awakening in the early 1800s. In the eastern U.S., where the gospel was spreading rapidly, another ideology began to emerge. In a counterfeit of God's outpouring of his spirit, some were turning to Spiritualism, a religion built around communication with the spirits of the deceased. Based on earlier works by "psychic scientists" Emanuel Swedenborg and Franz Mesmer, spiritualism taught that human spirits, when freed from the body gained great wisdom and abilities, and that these entities could be very beneficial if handled properly. Although using the methods of the typical psychic (seances, psychic readings, trances), spiritualists raised the practice to a religion.

Spiritualism eventually became a popular movement with around 8 million members by the beginning of the 20th century. Adherents were typically upper-class and included notables such as Arthur Conan Doyle, Thomas Edison, Mary Todd Lincoln, and Pierre and Marie Curie. Their "gospel" was spread mostly by lectures given by the more prominent members.

Spiritualism had a significant influence on 20th-century North America, informing many modern supernatural beliefs. Its concepts spread "innocently" yet stealthily through the entertainment media of the Victorian age through the early

1900s.[113] There are still practicing spiritualists today, however, the rhetoric has been toned down considerably, and most proponents work through established New Age and occult groups.

Witches, Warlocks, and Wicca

Witchcraft or simply the *Craft* is yet another ideology with roots beyond written history. As with most ancient belief systems, there are diverse forms under the same name, so someone will protest at any attempt to define it. One commonality among witchcraft groups is the practice of magic in all its variations—spells, necromancy, incantations, etc. Witches and warlocks (masculine) may employ *natural* magic forces or *personal* magic granted by supernatural entities. Generally, witchcraft is more on the practical side, not overly concerned with where powers are derived. Regardless, everything about witchcraft is condemned by Scripture…

> There shall not be found among you anyone who burns his son or his daughter as an offering, anyone who practices divination or tells fortunes or interprets omens, or a sorcerer or a charmer or a medium or a necromancer or one who inquires of the dead, for whoever does these things is an abomination to the LORD (Deuteronomy 18:10-12).

The media certainly makes the Craft seem alluring, and if rebellion or peer acceptance is the objective, sorcery is just the thing. In the widespread popularity of witchcraft, some question

[113] Some popular spiritualist fiction from the era: William Dean Howells, *The Undiscovered Country* (1880); Henry James, *The Bostonians* (1886); Bolesław Prus, *Pharaoh*, (1895); H.G. Wells, *Love and Mr. Lewisham*, (1900); Hamlin Garland, *Tyranny of the Dark* (1905); Arthur Conan Doyle, *The Land of Mist*, (1926); George S. Kaufman, *You Can't Take It with You* (1936,); Muriel Spark, *The Bachelors* (1960).

the "sincerity" of contemporary witches since so many, especially adolescents and teens, are caught up in the practice more as a fad. The common excuse is "They're not really into it—just curious." But exactly how closely must one follow Satan to be considered a follower? Satan is probably as pleased with witch and warlock wannabes as he is with the hardcore practitioners.

In the 1950s, a retired British government worker pulled together several ancient occult ideologies and created *Wicca.* Intended as a modernized, more "user-friendly" form of witchcraft, Wicca has evolved into many different subgroups called *traditions*. With few centralized elements, local *covens* within a tradition are free to "customize" their beliefs. New members must undergo ritual initiations with a coven and may rise to a third-degree priest or priestess.

Primarily a European and North American phenomenon, Wiccans may be polytheistic or pantheistic, however, most traditions tend to be *henotheistic*, with a male-female dual godhead. Most Wiccans describe themselves as *neo-pagans* rather than witches and deny any association with satanism. Their rituals center around pagan ideologies and involve equinoxes, solstices, lunar cycles, sexual perversion, and Samhain (Halloween). Wiccans take full advantage of modern social media with many apps, websites, and a robust cyber-presence. The dangers of Wicca are the same as for witchcraft in general, except perhaps they are more subtle. The Wicca Rede (creed), "If ye harm none, do what ye will" places a friendly façade on a belief system condemned by God.

Satanism

Perhaps the ultimate irony in all creation—worship of a being whose aim is the destruction of the worshipper. A.W.

Tozer taught that all false religions begin with an erroneous idea about God. The satanist, like Satan himself, believes the lie that God can be placed on an even plane with another. But then, *deceiver* is Satan's niche. It's what he does best. Thus, the follower of Satan sees him as an alternate, yet equally valid choice among deities.[114]

What does Satan think about satanism? Is it his ultimate goal to gain the worship of humans? I couldn't begin to guess the mind of the devil, but what is revealed of him in Scripture leads me to think this isn't the case. At some point in epochs past, glory may have been on his agenda, but for now, I believe his single-minded goal is to oppose God by deceiving, blinding, confusing, discrediting, and destroying his image-bearers in any way possible. Whatever best supports this objective will be his strategy. I believe Satan is just as pleased in trapping a Christian with pornography as he is in producing a satanic priest. He is equally satisfied leading those who believe he is a myth or those who worship him faithfully. Satan invented the term "result-oriented."

However, there remains among his schemes a place for those who choose to offer him worship and allegiance. Satanism began with the angelic rebellion in heaven, but the modern rebellion officially began with the founding of the Church of Satan (1966) and the publication of *The Satanic Bible* (1969). These were the work of Anton LaVey, a burlesque musician in San Francisco. LeVay cited his developing cynicism of religion, as he witnessed the hypocrisy of Christians attending his bawdy venues.

It's interesting that LaVey was an atheist, dismissing the idea of God, Satan, and the supernatural in general. For him, Satan was simply a good symbol for his rejection of religion. His teachings in *The Satanic Bible* were mostly hedonistic, promoting freedom from our guilt and inhibition over

---

[114] I differentiate between Satanists who believe Satan is opposed to God, and worshippers of Lucifer who see him as equally good or part of some pantheistic godhead (more of a New Age stance).

sensuality. LaVey's rock-star popularity, though, spawned a myriad of other satanic cults and ideologies. As with witchcraft, the attitude of "anything goes" permeates modern satanism, and any group can identify themselves as "satanic."

Satanists have been accused (at times rightly so) of animal mutilations, vandalism, arson, and even murder. The frenzy of satanic activities in the 80s and 90s was so widespread that many law enforcement agencies began to downplay or ignore the phenomenon. I would never attempt to excuse satanism, but any mass-murderer or pervert can *claim* to be a satanist. And again, sin is sin. *The appearance of a random babel of beliefs is no indication that a single-minded strategy isn't being carried out.*

We could study a long list of gruesome, evil, and paranormal accounts of satanism, but I don't believe the Enemy deserves the attention. We know his capabilities. For the Christian, perhaps our focus shouldn't be on analyzing and categorizing satanism, but in realizing who stands behind all sin and lawlessness. The Father of Lies can and will use organization, disorganization, fear, shock, normal, and paranormal, yet we have the promise "He who is in you is greater than he who is in the world" (1 John 4:4).

Halloween and Holy Days

I don't celebrate New Year's Day… Don't judge me… I have nothing against those who enjoy crowds in sub-zero wind watching a large ball drop. That's all three hours past my bedtime. And as a teacher, New Year's Day means vacation time is over and I have to rewrite several misdated checks. Other holidays are higher on my list—Christmas, Resurrection Day, our anniversary, my birthday, Independence Day, Memorial Day.

Theoretically, holidays (holy, "set apart" days) are designated as remembrances of something important. This has been the traditional Judeo-Christian view. Similar to holy *places*, holy *times* aren't inherently holy, rather their "set-apartness" points to the truth behind them. This is why Paul could advise…

> One person esteems one day as better than another, while another esteems all days alike. Each one should be fully convinced in his own mind (Romans 14:5).

How one acknowledges, obeys, and conveys godliness to others is more important than the particular date. Yes, God does command to acknowledge certain days (*e.g.* the Sabbath), but the emphasis is always on the truth behind the act, not the innate value of the time period. This isn't the view of many pagan holy days.

The tradition of Halloween is an example. To the Celtic pagans (my ancestors), October 31 (*Samhain*, pronounced *sau-wen*) was the last day of summer. For them, winter began on November 1 as well as the solemn preparation for the approaching cold, dark, dangerous days with no fresh vegetables. In addition to physical stress, Samhain was believed to be a time in the solar cycle at which the veil between the natural and spirit realms grew thinner. Portals would open and the spirits of the dead, along with demons, could walk the earth for a short time.

Samhain night was *All Hallows Evening*, a time to remember, respect, and often to appease spirits. You never knew which ghost might come to visit. It could have been Uncle Angus whom you accidentally ran over with your oxcart. To protect your family, you could build a big fire, put spooky carved gourds around your house, or disguise your family as ghouls and goblins.

Times were hard for the Celts and there apparently were lots of miffed ancestors and angry dead townsfolk to contend with on Halloween. Often, in the confusion of the night, all manner of evil was tolerable. It was a good time to knock off your irritating neighbor who played his bagpipes late at night. Just blame it on a ghost. Sexual perversion was common, since apparently devils can make you do just about anything, and it isn't your fault.

In the 17th century, Pope Gregory decided enough with demonic holidays in the empire. To avoid messy skirmishes, he instructed missionaries to keep the holy days, but try to put a Christian spin on them. So, along with many other pagan holy days and rituals, Halloween was "Christianized." The holiday was now All Saints Day (Nov. 1), a time to respect believers who had passed on. Scary stuff was still used to keep demons from spoiling the festivities. The unbelievers, however, maintained the old ideology.

Eventually, the day became All Souls Day (Nov. 2), a time to pray dead saints out of purgatory. Needless to say, Luther and his cohorts didn't like that idea. The Protestants taught that all pagan practices were anathema. As for North America, the Puritans and most other colonists didn't celebrate Halloween (or Christmas). All Souls Day was brought to America by Catholic groups. Most of our modern Halloween customs and iconology came with the Irish-Scottish migrations of the 1800s.

A puzzling relationship exists between many Bible-believing Christians and Halloween. As a child, we celebrated Halloween. I enjoyed scaring people, especially girls. I also remember a twinge of cognitive dissonance when I reflected on ghouls, witches, skeletons, death, and evil in general, but hey, it was just for fun!

If I had a choice, I think I might erase Halloween from the calendar. When I consider the reality of Satan and the devastation he causes—disease, suffering, hatred, sin, death—it's difficult to justify frivolous representations of his domain, regardless of how innocent we attempt to make them. We don't have to search very far in Scripture to find commands to "abstain from every appearance of evil."

Yet I realize other faithful believers disagree with my trepidations. Some suggest that Christians should actively "take back what Satan has claimed and sanctify it for the Kingdom." There is some truth in this, but can something be so corrupted that we should just trash it all and start over with something new?

Others claim that Halloween celebrations are a good way to evangelize unbelievers. I've seen this work, but only when the iconology doesn't contradict the gospel it's intended to convey. I question the effectiveness of presenting Christ's love while ghosts are chainsawing zombies in the background.

As Paul instructed, the way in which holidays are celebrated should never be a dividing issue among believers. We are, however, responsible for our decisions, especially their effects on others. Our ultimate goal is always, "Whatever you do, do all to the glory of God" (1 Corinthians 10:31).

Voodoo and Santeria

I've included these in this chapter since they are culturally associated with the occult. Both Voodoo and Santeria are considered religions or at least belief systems. They are similar in many ways, Voodoo having French origins and Santeria Spanish. This discussion will be very general since both beliefs have little written, organized tenets.

These are *syncretistic* religions, meaning the beliefs are a mixture of ideas from other religions. Voodoo and Santeria developed as African slaves were brought to the islands southeast of the U.S., where their African polytheism merged with mostly Catholic Christianity. They are considered monotheistic now since both profess a single chief deity. Contact with the divine, though, is accomplished through lesser deities which are believed to be angels, demons, or dead humans. The Catholic concept of saints is an easy fit for followers.

Voodoo and Santeria are highly individual and experiential, so there is a broad range of associated beliefs and practices. New revelation can come to any practitioner, and communities may have their own unique tenets. Members practice many forms of the supernatural, including divination, magic, spells, curses, and omens. The voodoo doll or effigy is a well-known aspect of syncretic magic. Animal sacrifices are common in many communities. In several sects, the worship of Satan is accepted since he is considered one of the sub-gods and is equally able to cooperate with those who summon him.

It isn't difficult to see that Voodoo and Santeria believers don't worship Yahweh God of Scripture. Practitioners actively seek spirit engagement; thus, manifestations of demonic activity are common. There is a trend among young people to dabble in the mysteries of the two religions. However, for the Christian, God's admonitions against magic and divination are unwavering.

10
Bigfoot and Friends

The account of trapper Bauman from Theodore Roosevelt's book *The Wilderness Hunter*...

About midnight the thing came down through the forest opposite, across the brook, and stayed there on the hillside for nearly an hour. They could hear the branches crackle as it moved about, and several times it uttered a harsh, grating, long-drawn moan, a peculiarly sinister sound. Yet it did not venture near the fire...

[The next morning] Bauman could see nobody; nor did he receive an answer to his call. Stepping forward he again shouted, and as he did so his eye fell on the body of his friend, stretched beside the trunk of a great fallen spruce. The footprints of the unknown beast-creature, printed deep in the soft soil, told the whole story.... While thus waiting, his monstrous assailant, which must have been lurking in the woods, waiting for a chance to catch one of the adventurers unprepared, came silently up from behind, walking with long noiseless steps and seemingly still on two legs... It had not eaten the body, but apparently had romped and gamboled around it in uncouth, ferocious glee, occasionally rolling over and over it;

> *and had then fled back into the soundless depths of the woods.*[115]

With place and name changes, tens-of-thousands of similar (but not as deadly) incidents have been reported in the US. Bigfoot creatures are the "poster-monsters" in a larger group of mysterious alleged creatures known as *cryptids* or "hidden animals." Those who research these alleged unusual species take the title of *cryptozoologist*.

Of course, the existence of strange, undocumented animals doesn't inform our worldview to the degree of other unusual phenomena, nor does it necessarily rise to the level of paranormal. However, an overview of cryptids provides a good mental exercise. Many of these same principles can be applied as we look at other phenomena. Also, there are important connections to the Christian worldview related to certain alleged cryptids. Furthermore, Bigfoot and other cryptids just can't seem to keep their noses out of the paranormal—at least in pop culture. We'll examine these more bizarre ideas in the next chapter.

"Unusual" is certainly a relative term when speaking of earth creatures. We are blessed with about 8 million different species right now, not counting many more that are extinct. When I thumb through any children's animal book, I'm still amazed by some new creature of which I've never heard.

Sightings of *very* unusual animals are part of our folklore heritage. We can assume that a large portion of these might be misidentifications and exaggerations, although they are likely based on some facts. I can imagine a drunk sailor mistaking a sirenian for a mermaid. Things chasing you always look bigger and fiercer in the rear-view mirror.

---

[115] Theodore Roosevelt, *The Wilderness Hunter*, (New York: G.P. Putnam's Sons, 1893), 258-261.

Cryptozoology at Work

I've spent a lot of time admiring the awesome diversity of plants and animals on our little planet. In all my trekking, I've yet to discover a new species. Cryptozoologists (CZs) are typically pushed to the fringes of mainstream biology, but I certainly understand their passion. It would be amazing to find an undocumented species of daisy, but imagine coming across a strange, new large mammal. It makes me want to go Sasquatch hunting!

Perhaps the most significant difference between CZs and general zoologists is in their methodologies. Field zoologists usually search for new species by conducting a broad survey of a habitat with eyes open for anything new and interesting.[116] CZs, however, use *targeted* searches for a particular species that has been reported anecdotally.[117] This is still a valid method, given the reliability of the initial observations. In fact, it isn't uncommon for a conventional zoologist to narrow a search to a specific animal or plant that someone might have reported.

Classified Cryptids

It's unfortunate to see all cryptids lumped together and dismissed as a group. Here again, we see an example of a kind of *composition fallacy*. There's a range of "believability" among cryptids, and it wouldn't be good science to judge all by a few (or *vice versa*). As with other phenomena, reports are mountainous, and it helps to look at the data in their proper categories. I'll use a system with four *types* based on Arment...[118]

---

[116] Sometimes call a bio-blitz, many people sweep through an area and record every species they find, sometimes collecting a new species.

[117] Cryptids are said to be *ethnoknown*, as opposed to species which have already been documented.

[118] Arment, *Cryptozoology*, 16.

Type 1 Cryptids... Organisms similar to living, documented species but with significant physical differences. Some new species are classified as *cryptic species* (not to be confused with *cryptids*).[119] These are populations that look so much like another species, no one ever noticed that they were different. For example, it was discovered in 2014 that there could be several different species within the single species we now call giraffes. This was a highly technical determination, based on mitochondrial DNA, so it was difficult to tell just by looking.

Would a Bigfoot, by its composite description, fit into this category? Are they simply large primates similar to other large primates except taller and more bipedal?

Type 2 Cryptids... Organisms similar to living, documented species, but "out of place" geographically. A common cryptid story from the southern U.S. is the existence of large black "panthers."[120] This cryptid has been reported for decades but are assumed to be misidentified feral cats. There are several habitats with cougars (mountain lions, pumas), although a true black specimen has never been documented. We now know that jaguars do range into the U.S. from Central America.[121] We also know that black *sports* of the animal are part of the population. It wouldn't be unheard of that some of the millions of people in the southern U.S. might encounter a black jaguar.

Also, the possibility of feral animals could dispel some of the mysteries. I know of at least a dozen local individuals who raise exotic species as a hobby, some of which would make headlines if encountered in the local woods. A group of hominid

---

[119] Although is this a real difference? Had someone reported that a new species of giraffe might exist, before the difference was found, it would have been a cryptid.

[120] I don't have many of my own anecdotal tales of the "paranormal" except that I did see a large black cat in 1972. It crossed the road in front of me in an isolated area late one night. The cat was as long as my VW and was totally black I got a good look from thirty yards, and as others have said so scientifically, "I know what I saw!"

[121] Hilary Hanson, "One of the Only Wild Jaguars Known to Roam the U.S. Is Believed to Have Been Killed," www.huffpost.com, June 23, 2018.

sightings in Florida seems to indicate an animal similar to an orangutan. Could cryptids such as Bigfoot fit within this category? If a Sasquatch was seen in the Congo, might it be interpreted differently?

Type 3 Cryptids... Organisms similar only to extinct documented species. Recently, sightings of thylacines (Tasmanian tigers) have been increasing "down under." Thylacines were one of the largest predatory marsupials to stalk Australia and Tasmania. In the late 1800s, the government placed a bounty on the animals due to their threat to livestock. The last living animal was documented in 1933. Some zoologists accept the possibility that a breeding population could be surviving in remote areas. With numerous recent reports of thylacines, especially in Tasmania, it would seem the existence of this species isn't as improbable as some other cryptids.

There is precedent for fossil species to be discovered extant. In 1938 a fisherman's catch included a fish that had been previously identified through its fossils. Thought to have become extinct 60 million years ago, the lobbed-fin coelacanth was discovered off the coast of Africa. This situation does raise the question as to whether the potential existence of other extinct species is only a quantitative difference.

Is the length of time from extinction to the present a significant factor in evaluating the tenability of a particular cryptid? The age of the earth (universe) is an in-house discussion among Christian apologists and scientists. There is a consensus *that* God created. The debate is over *how*. Briefly, old-earth (long-day) proponents maintain that the universe is the billions-of-years-old that most secular naturalists claim. An old-earther would say that God has used evolution over the eons to bring the creation to its current state, and the "days" mentioned in Genesis 1 metaphorically refer to long time spans. Young-earth (short-day) creationists believe that God created the universe in the six, literal twenty-four-hour days as described in Genesis 1. Thus, young-earthers see the earth as 10,000-years-old, give or take.

I won't argue one way or another here, except to say that the two views offer different starting assumptions regarding some cryptids. For example, some would entertain the possibility that *mokele-mbembe*, a creature reported in the Congo, is a small quadruped dinosaur in the brachiosaur group. Its composite description is essentially that of the brachiosaur species, for which we have ample fossil remains. No one doubts that these animals existed in the past. Mainstream science and old-earthers, however, have problems with the idea of extant dinosaurs. The last of the dinosaurs supposedly became extinct sixty million years ago. The chance that a particular species (as a larger breeding population) could still be alive after millions of years would be considered very slim.

But the possibility increases significantly if the period is shortened to only thousands of years. In fact, a literal, young-earth view would place humans with dinosaurs in the original creation anyway.[122] Old earth Christians, as well as secular scientists, find this notion unpalatable. The standard complaint is that a literal six-day creation would require too many events beyond our current observations of the physical world (miracles?).

Many scientists use a timeline, based on billions of years of evolution, and many solid Christians accept this view. Science is rightly established on the *principle of uniformity*, which states that "the present is the key to the past." In other words, the most logical reasoning begins with discovering how the present world works. We then extrapolate back into history.

---

[122] There is no doubt that huge, scary, reptilian monsters have roamed the earth. I stand in awe when visiting the forty-foot-long, *Acrocanthosaurus atokensis* skeleton at the Museum of the Red River in Idabel, Oklahoma. I try to imagine this creature stomping through my backyard. The visual effects on Jurassic Park really drove home how real these creatures were. The bottom line is, fossil remains of dinosaurs and many other creatures provide the only data we have on these animals. All else rests on interpretation.

This is, indeed, good science. It becomes problematic, though, when it leads to the exclusion of events that we don't observe in the present or recorded history.[123]

The issue comes up when considering the biblical account of a worldwide flood. Since we've never witnessed such an event, geologists tend to assume that the earth's features must be explained by processes active today. For example, the current processes of erosion could have produced the Grand Canyon, given enough time. But wouldn't a major, "non-uniform" event be able to accomplish the same erosion in a shorter time? Once again, our ultimate answer might depend on *whether our worldview accommodates miracles and the supernatural.* The bottom line for our present discussion is that the existence of these paleocryptids isn't the problem for young-earth creationists that it is for proponents of a vastly older universe.

Type 4 Cryptids... Organisms that are not similar to any single known species, living or extinct. In my opinion, this category is a no-brainer. I enjoy nature series and documentaries, but they usually give skewed views of earth life. In an effort to showcase the diversity of species and animal oddities, they often skip over the commonalities and patterns among the totality of species: Vertebrates are based on a four-leg design... Out of thousands of fish species, they all are variations of the same fish body... All one-cell organisms have virtually the same parts... It's *extremely* uncommon to see a species with features *totally* different from any other.[124]

What about animals with a combination of physical structures from several species? Some animals seem to have

---

[123] The philosophical debate is between *uniformity* and *catastrophism* (the term here means "out of the ordinary", not necessarily destruction). Young-earthers agree that uniformity is a good general principle, but they accept at least three major catastrophes in earth's history—the creation, the fall, and the flood. Each of these involves supernatural elements.

[124] Evolutionists might attribute this to a common lineage—Creationists to a common Creator.

features of another group. For instance, the platypus seems to have a beaver's tail, a duck's bill, and a chicken's egg-layer. Closer examination, however, shows that no rules were broken and that this composite view is just a human artifact. Other related mammals can produce eggs (different from bird eggs), just like the platypus. Several aquatic mammals possess flat tails. And that duck bill—more like hairless lips than a bird beak. As weird as it seems, the platypus is just another variation on an earth-animal theme.

I think, however, the believability line is crossed when *chimera*-type cryptids are suggested. A chimera is a proposed composite of two or more unrelated animal *forms*. In Greek mythology, the original Chimera had a lion's body with an actual goat head protruding from its back and a snake for a tail. What moves this chimera to the crypto-trashcan? It is not that this is just totally weird. Simply, *the creature doesn't fit the pattern of an earth animal.* A real goat head and a real snake fused onto a lion is inconsistent with the rest of zoological anatomy.[125]

Throughout history, much folklore has been based on chimeric creatures. The discovery of a living unicorn would thrill a lot of people and pump millions in revenue into the notebook/backpack industry, but would it be impossible? Perhaps not. If a unicorn is a horse with a horn, we already have them in Africa. Rhinos are in the same order as horses. They clearly have a type of horn and a "horse-ish" form. Even better, a fossil form more resembling an actual horned horse was recently discovered in Siberia.[126]

There are eight references to unicorns (*remim*) in the Old Testament. They are mentioned in lists with other known

---

[125] Although, in a later chapter we'll see how modern science may twist this rule of chimeras.

[126] Andrei Shpansky, Valentina Aliyassova, and Svetlana Ilyina, "The Quaternary Mammals from Kozhamzhar Locality," *American Journal of Applied Sciences*, October 2, 2016.

animals, so the argument isn't whether they existed at the time but to the specific animal in the mind of the writer. Various translations offer them as wild oxen, antelopes, or other horned beasts. Some contend that the biblical unicorn was a rhino, but why make up a new name for the known animal? From living types and fossils, we know of at least 50 different species of rhino-like creatures, some more like horses than the modern ones. I'll side with every adolescent girl in the world and say it—there were unicorns at one time. Furthermore, I would be highly intrigued by the discovery of a living unicorn, but it wouldn't alter my worldview.

Ah, but let's add feathered bird wings to a horse. Still possible? I have a problem with Pegasus. Yes, birds have bird wings, but they're not mammals. Bats are mammals and they have wings. Could Pegasus be like an oversized, mammalian bat? There's that earth-pattern problem again. Four-appendage-flyers exist, but two wings on Pegasus plus four legs equal six appendages. Six-legged insects are okay, but the earth-pattern doesn't support six-appendage vertebrates—period. (Let's not even start on the feathers.)

Jackalopes? An antlered bunny? I have my doubts—especially since no hint of headgear is known for that species. Merfolk? The anatomical awkwardness alone can rule them out. Faeries? Tiny, cute people, a very distant maybe, but definitely not with two extra wing-appendages. I believe the improbabilities of Type 4 cryptids are intuitive to most people—at least to adults.[127]

---

[127] Interestingly though, chimeras also play into modern genetic technology. It surprises most people to learn that we have produced many chimeras in the past decade—even some human/animal combinations. Although it sounds like science fiction, these chimeras are often used in researching human transplant options. During very early embryonic stages, cells from different species may be combined to produce a new organism with cells from both donors. In one project, human cells were combined with sheep cells and successfully grown to maturity. The technology is limited only to the imagination of the scientists and laws that might add restrictions. At some time in the future, Pegasus might be an option.

Are There Undiscovered Species?

Cryptologists often pick on paleontologist George Cuvier, who announced in 1812 "There is little hope of discovering new species of large quadrupeds."[128] Within a few years, explorers found gorillas, pygmy hippos, giant pandas, Komodo dragons, and several new antelope species. So, he was a bit premature in his prediction, but 200 years later—surely we've found all there is to find. Perhaps not.

Between 1993 and 2008, zoologists discovered 408 new species of mammals—10% of the already known species.[129] As expected, new rodent species made up 174 of these, but there were larger mammals as well. There were 11 new hooved mammals documented. Surprisingly, 55 new species of primates were on the list. Of the total mammals, 55 families added a new genus.

In 2003, zoologists were surprised to discover a population of new primates in the jungles of Tanzania. This new species *kipunji*, is related to baboons. Regarding the discovery of kipunji, *Smithsonian Magazine* deemed it…

> …astonishing for a large primate to be discovered in the 21st century in a heavily populated corner of East Africa, where human beings have been kicking around for as long as we have been human.[130]

---

[128] Martin Rudwick, *Georges Cuvier, Fossil Bones, and Geological Catastrophes: New Translations and Interpretations of the Primary Texts*, (University of Chicago Press, 2008).

[129] Gerardo Cebados & Paul R. Ehrlich, "Discoveries of New Mammal Species and Their Implications for Conservation and Ecosystem Services," *Proceedings of the National Academy of Sciences of the United States of America,* 2009, 106(10), 3841-3846.

[130] Richard Conniff, "Meet the New Species," *Smithsonian Magazine,* www.smithsonianmag.com, August 2010.

Although not as well-publicized as discoveries of the past, field zoologists are currently cataloging more species from more habitats than ever before in what some call a "new age of discovery."[131]

Most of the undiscovered animals in the world obviously aren't mysterious sea giants or hulking ape creatures. The majority (as they are discovered) will likely be microorganisms, insects, small mammals, and such. In fairness to the cryptozoologists, this is one reason cryptids are often sensationalized. A hiker chancing upon an undocumented mosquito species isn't likely to discern the difference, even on close inspection, thus this species will never be noted as a "cryptid." But clearly, a seven-foot hairy humanoid running across the trail might stand out as something unique. Cryptids by definition must be conspicuous enough to deserve observation and subsequent storytelling.

A Closer Look at Bigfoot

Accounts of Bigfoot and other cryptohominids such as Yeti, skunk ape, hairy man, grass man, and Sasquatch date back at least hundreds of years and seem endemic. About a third of the 23,000 sightings in North America are concentrated in the Pacific Northwest. The rest are dispersed in clusters through forested and wetland areas. Descriptions vary somewhat from one "community" to another, yet individual sightings in a particular locale seem to be consistent.

Cryptid investigation generally starts with two questions:

- Could this creature exist?
- Could this creature exist *here*?

[131] M. Donoghue & W. Alverson, "A New Age of Discovery," *Annals of the Missouri Botanical Garden,* 2000, *87*(1), 110-126.

Could a creature such as Bigfoot exist? I like to use the "what-if-I'd-never-seen" test on cryptozoological icons. For example, what-if-I'd-never-seen a platypus? Can you imagine trying to convince your friends that you saw a hairy mammal with a duck's bill and venom claws, and it laid eggs like a bird?

What-if-I'd-never-seen a gorilla? I must admit, gorillas are awesome, especially if I happened to meet one in the woods and I had never heard of such.[132] My description of the creature would definitely arouse the cryptozoologists. But are gorillas qualitatively different from Bigfoot? Except for the feet and a few other deviations, not so much. Objections to cryptohominids are based more on *time and geography* than on *morphology* (a Type 1 cryptid).

Think of an encyclopedia page picturing the range of families within the primate order. Would an image of a Bigfoot on the page seem inappropriate to an unbiased observer? Primate researcher Jane Goodall doesn't think so…"Well, now, you'll be amazed when I tell you that I'm sure [Bigfoots] exist."[133] Her view is based on her expertise with primates and the assumption that a Bigfoot is a primate.

Most honest skeptics don't oppose the notion that a cryptohominid *could* exist in North America. Instead, the rational argument might be that given…

- limited territory,
- a limited number of sightings (considering the large population of people over many of years),
- that we have lots of sightings of other animals (even big ones),
- the lack of convincing physical evidence (photos, tracks),

[132] Gorillas were identified relatively recently (1847). Prior to that, tales about cryptid "ape-men" in the area were common.
[133] September 27, 2002, NPR, *Talk of the Nation: Science Friday*.

- that a cryptid would likely mean a group of cryptids…

…*does* a cryptohominid in this area seem likely?

The disagreement will always be somewhat skewed, however. Science is much better at *falsifying* than *verifying*. Skeptics must disprove each bit of evidence found. "Believers" in Bigfoot need only produce one undisputable bit of evidence. But the onus is on the believer to find good evidence. Every piece of unproductive evidence adds to the weight of the skeptic's position, as does the passage of time.

So, Bigfoot—thumbs up or down? I remain somewhat uncommitted, leaning slightly toward "yes." There is no reason that a large primate *couldn't* live in North America. To some, the preponderance of sightings in the northwest could be explained by the ice bridge that once existed between northern Asia and Alaska, allowing migration of Asian primates.[134] This could provide a genetic link to the alleged Yetis and known large primates.

If an American primate exists, it would probably look somewhat different from known primates, and it would be adapted to its particular ecosystem. It would have to have a breeding population (unless it's the last of its kind and on the way to extinction), and that would require a large territory. Extremely large areas can be found in North America. When I fly over the boundaries of southern Arkansas and Oklahoma, I'm impressed by the hundreds of square miles of state park and forestry land, with only sporadic human activity. Some areas are so dense and inaccessible that one could easily imagine a group of secretive animals that would rarely have contact with humans. A fact that adds a degree of validity to Bigfoot

---

[134] Fossils of 9-foot *Gigantopithecus*, a hominid similar to an orangutan in body structure, have been discovered in Asia.

sightings is that the majority are in coniferous forests.[135] Cryptozoologists speculate that conifers may provide food (high-protein conifer seeds) or year-round canopy cover.[136]

Sasquatch by Any Other Name

Legends of Bigfoot-type creatures have arisen independently in many cultures. The descriptions seem to be variations on a theme. Is this evidence for a real animal or simply our shared archetypes?

- Orang-Pendek – Sumatra.
- Yowie – Australia.
- Grassman – Ohio.
- Skunk Ape – Southeastern U.S.
- Barmanu – Pakistan.
- Momo – Missouri near the MS River basin.
- Yeti – Himalayan Mountains.
- Urayuli – Alaska.
- Chuchuna – Russia.
- Fouke Monster – Southwestern Arkansas
- Mapinguari – South America.
- Oily Man – Malaysia.

---

[135] Proponents claim that if reports are hoaxes or mistaken identity, we would expect sightings to be equally spread over all forest types.

[136] I've pondered the possibility of large hominids in my area (southern OK and AR). I tend to make comparisons with our bear population, estimated to be around 2000 (about one per 5 sq.mi.). Obviously, there is habitat and food for this population. Most CZs assume Bigfoot would have similar omnivore diets. Bear encounters are relatively rare in this territory. If a hominid population of, let's say, 20 existed here, we might extrapolate an extremely low incidence of encounter. My point is, the idea of a Bigfoot group in the area wouldn't be "outrageous." There are still many unknown variables: shared niches, competition with bears, hibernation, avoidance of humans.

Bigfoot has unquestionably become a folk hero across North America. It would be naïve to think hoaxes aren't some part of the phenomenon. We know some incidents have been staged. Once again, it would be equally naïve to designate all Bigfoot evidence as contrived. The range and nature of the surroundings at the sites seem to speak against this as an all-inclusive explanation. In my locale, hoaxers tramping through the woods in a hairy suit would quickly become rarer than most cryptids.

We might also predict that as the human population increases and habitat decreases, the likelihood of documenting a cryptid increases. Proponents argue that that is exactly what is occurring. Sightings are increasing due to expanding human activity. Of course, a cryptid can never be scientifically verified without a body or a genuine DNA sample.

A Closer Look at Sea Monsters

Sea monster stories have been around as long as there have been sailors. It's easy to blame the stories on long periods of isolation, monotony, or a bottle of rum. But the ocean is a really big place. We landlubbers forget the immensity of the ocean habitat. Almost 70% of the earth's surface is ocean, with a total volume that is staggering. If mysterious creatures do exist in the ocean, they have over 300 million cubic miles in which to hide. The deepest regions are over six miles below the surface—far beyond our present abilities to explore adequately.

And admittedly, *monster* is a difficult term to nail down. What-if-I'd-never-seen a whale, for example? I can't imagine a more stereotypical "monster." An animal that could wreck a ship with one tail flip, then swallow the crew![137] Squids and octopi are even scarier. We now know of one species of squid that grows to almost fifty feet, making it a perfect candidate for

[137] Although most species are plankton feeders.

a monster movie or the fearsome Kraken. If perchance I ever encounter a sea monster, would it be qualitatively different from these?

We seem to have become "monstered-out" on those denizens of the deep, though. In the past century, most of our folklore has turned toward smaller lake monsters. Since few of us venture onto the high seas often, lake-shore picnics provide a much better opportunity to spot an aquatic (freshwater) beast.

The most popular aquatic cryptid is undoubtedly the Loch Ness Monster, or Nessie. With 99% Scottish blood coursing through my veins, my official stance has to be that Nessie is real. Unofficially, however, I'm allowed to ask questions.

Is Nessie a Dinosaur?

For all practical purposes, Nessie's descriptions are those of the dinosaur *plesiosaur*. The unspoken issue for many scientists is that the existence of a modern-day dinosaur would contradict evolutionary and old-earth systems of dating.[138]

But if we set aside the dating problems, is it even possible for large water cryptids to survive in the places reported? Here again, habitat and potential for human interaction are major factors. Lakes are *much* smaller in volume than the ocean. Some claim that Loch Ness has features that make it a suitable location for the monster. It does have an outlet to the North Sea. The possibility has been offered that the lake might simply be something of a breeding area for the creatures that normally hide in the depths of the ocean. That would explain the inability to corner Nessie with very intense searches. Also, the extended range could provide the required habitat. A famous North

[138] There is also the ongoing problem of other large, modern reptiles. Some zoologists argue fine points of bone structure, phylogeny, and other anatomical variations, but crocodiles, Komodo dragons, and extant reptiles in general are not significantly different from dinosaurs in the context of this discussion.

American lake cryptid reported in Lake Champlain seems to share some of Nessie's traits. The long-neck reptilian creature, called Champ, is a significant economic asset to the New York/Vermont region.

The possible existence of Nessie, Champ, and the mokele-mbembe raises some interesting questions. As mentioned, modern dinosaurs would not be the problem for young-earthers that they are for old-earthers and evolutionists. A recent creation (and the recent extinction of most dinosaurs) would be more consistent with a literal interpretation of Scripture.[139]

Some point to the unusual beasts in Job 40-41 as evidence that dinosaurs and humans were contemporaries. To show Job that he was truly the Almighty, in perfect control of the universe, God displays two of his most magnificent creatures. I believe it's noteworthy that God devotes 44 verses of his inspired Word to describing the animals. The implied intent was to overwhelm Job's doubting with clear evidence of God's sovereignty. Naturally, the examples would have to meet the expectation of awe and wonder. This is where I think some interpretations miss the mark.

In many translations and footnotes, *behemoth* (similar to the general term for "beast," or "cattle.") is identified as a hippopotamus or elephant. I like hippos and elephants. They're interesting, photogenic, and sometimes quite amusing, however, I wouldn't put them at the level of overwhelmingly awesome.[140] I recall vividly my first viewing of the original *Jurassic Park.* My expectations were somewhat higher than the dino movies of the 60s, but I must say the first view of that brachiosaur rising on its hind legs to eat a tree—now that was awesome!

---

[139] Information about the young-earth view can be found at www.icr.org and www.answersingenesis.org.

[140] Some claim that this is an example of biblical poetic exaggeration. This is employed in some books, but there seems to be no "poetic license" with the long description of other animals leading up to this. There's no indication that the mode changes with these final descriptions.

I like to imagine that Job may have never seen a behemoth (whose description fits perfectly with that of a brachiosaur). His reaction was probably similar to those of characters Sattler and Grant—utter shock. Sure, elephants are splendid creatures, but the description of behemoth seems very un-elephant: can't be tamed, semi-aquatic, tail like a cedar tree…

And if behemoth weren't enough, God calls up the sea-dwelling *leviathan*. The translators' best guess at this creature is usually a crocodile. Once again, I just don't see a croc in the description. Its features put even behemoth to shame…

- Can't be speared, hooked, bound, or shot with arrows.
- No hope of subduing it.
- No one able to fight it.
- Can't be skinned.
- Chest like stone.
- Treats brass like straw.
- Stirs the sea like a boiling cauldron.
- Nothing on earth its equal.

Ancients were familiar with some pretty fierce whale species. The verses imply that this beast was even fiercer. Some say the description sounds more like something of the mosasaur type. But here's the embarrassing little problem for those who prefer to limit God's ways to the modern, naturalistic *status quo*. Leviathan was a fire-breather…

> His sneezings flash forth light, and his eyes are like the eyelids of the dawn. Out of his mouth go flaming torches; sparks of fire leap forth. Out of his nostrils comes forth smoke, as from a boiling pot and burning rushes. His breath kindles coals, and a flame comes forth from his mouth (Job 41:18-21.)

Fire-breathing dragons have been too over-the-top for most theologians the past two centuries.[141] Excuses include metaphor, poetry, an orangish tongue, just really warm breath, mythology, and mistakes in Scripture. Some scientists suggest ways that such a feature could be plausible. I don't pretend to have the final word on these beasts, but what I won't do is limit my options to appease those who limit God and his Word simply because their naturalistic presumptions are rustled.

We would do well to remember the scope of the Book of Job. The first chapter begins with a supernatural council in the divine realm employing supernatural interventions into the earthly realm. The main theme is that *we don't always see the full supernatural picture, but God is in control anyway.* Moreover, God's final dialog wasn't intended to give Job the answer to every mystery, but to lead him to confess: "I know that you can do all things; no purpose of yours can be thwarted."[142] *If* the creatures in Job were actually dinosaurs, a serious blow has been made to "millions-of-years" theories. With such high stakes, it's no wonder the idea is controversial.

Nearing Trail's End

Many CZs believe we are at a crucial point regarding the "famous" cryptids such as Bigfoot and Nessie. The technology to "fake" physical evidence and videos makes the data inconclusive regardless of the quality. Only a body (dead or alive) or a unique DNA sample will verify their existence. But with the spread of other technologies such as drones, robotics, and IR photography, if a cryptid does exist, its days of secrecy are probably numbered.

Should a long-sought cryptid be documented by mainstream science, it would no doubt make headlines.

---

[141] There are numerous examples of animals with the ability to combine secretions and create light, heat, and electricity.
[142] Job 42:2

Ultimately, however, the creature would join the ranks of the rhino, gorilla, and coelacanth. When the mystique subsides, it will likely be added to the endangered species list, then tagged, tracked, and analyzed. Instead of a monster, we'll find another of the Creator's handiworks, perfectly designed to fit into an earth ecosystem.

11
Bigfoot on the Fringe

"I think [Bigfoot] is a kind of husband of Unk-ksa, the earth... Sometimes we say this One is kind of a reptile from ancient times who can take a big hairy form. I think he can also change into a coyote."
Joe Fly By[143]

"Paranormal people like Sasquatch and ET-types have social contact with humans in this world when they choose to be seen; otherwise, they go about their business." Kewaunee Lapseritis[144]

Once I had a bright idea for teaching seventh graders about moon phases. I brought a package of large round cookies to class. As I described what the various phases looked like, I gnawed the cookies into the basic shapes. There was a "no food" rule in force and I was instantly a rogue hero—but it was for science. I was so proud of my demo until exam day…

> "A half-moon occurs when gravity tears out one side of the moon…"

---

[143] Quoted in Keller & Knapp, *Hunt for Skinwalker*, (NY: Paraview, 2005), 158.
[144] Kewaunee Lapseritis, *The Psychic Sasquatch and their UFO Connection*, (Blue Water Publishing, 1998)

> "In the lunar cycle, sometimes pieces of rock fall away leaving a crescent shape…"
> "Phases are when part of the moon is like, eaten away…"

This is a dangerous chapter. Good teachers warn never to associate two superficially similar yet unrelated (non-analogous) ideas. Simply referencing ideas in the same lesson may indelibly link them in the minds of students.

However, my strategy here is strictly damage control. The deed is done. I believe the idea of true *biological cryptids* has been so merged with *paranoid cryptids* that our considerations are hopelessly confused.[145] Invariably, most *History* and *Trvl* channel features about cryptids begin with flesh-and-bone beasts and end up somewhere in the twilight zone…

> Popular interest in mysterious phenomena is high, creating an image problem for cryptozoology. Cryptids are often sandwiched between ghosts and UFOs within documentary-style programming, creating the perception that they are unearthly or paranormal… The need for entertaining or suspenseful devices creates the false impression that cryptozoology is just about hunting monsters…[146]

I urge the reader to maintain a clear separation as we look at biological and paranormal cryptids. A corollary to the Pickle Principle might be appropriate: *The existence or non-existence of one group has no bearing on the existence or non-existence of the other*. A biological cryptid may be reasonable—a paranormal cryptid is a whole different animal.

---

[145] Even referring to both groups as cryptids is probably misleading, but I'll do so here for simplicity.
[146] Chad Arment, *Cryptozoology: Science and Speculation,* (Landisville, PA: Coachwhip Publications, 2004), 11.

As previously stated, cryptids in and of themselves aren't paranormal. However, Bigfoot invariably takes on quite a different persona. "Parasquatch" has moved far beyond a flesh-and-bone primate. Later, we'll explore some links to other paranormal phenomena. For now, let's consider the supernatural side of Bigfoot encounters.

Obviously, with the popularity of Bigfoot and the large number of sightings, a few outlandish outliers could be expected. However, a growing number of reports involve abilities beyond a natural wild animal. Several Native American groups, for example, consider Bigfoot and kin to be spiritual phenomena. Depending on the specific tribal tradition, the creature's role ranges from a sacred and revered wild animal to a wise and powerful spirit entity.

Since the earliest modern sightings, some have sided with the tribes who view Bigfoot as highly intelligent—perhaps superior to humans. Although based on the *absence* of evidence, they point to the fact that Bigfoot has successfully eluded us for centuries. If Bigfoot is real, he certainly has "outwitted, outlasted, and outplayed" us—the less-hairy primates—always staying one step ahead. Proponents say that would be expected from an intelligent being, well-camouflaged and on his own turf. Could there be an advanced civilization of the creatures who prefer to live in secluded underground habitats (and who also prefer not to bathe)? Some have suggested Sasquatch is actually studying us. Several sqautch-hunters have begun wearing backward-facing cameras for that very reason. As Keel might say: "You noticed them, and they noticed that you noticed them."

Kicking it up a notch paranormally, there have been consistent reports of Bigfoot fading in and out of sight, similar to ghostly apparitions. There are stories of the beast disappearing into a cliff face or materializing in a beam of light. Those who accept the incidents theorize that either the creatures are “cloaking” themselves in some way or are moving through dimensional portals. Both of these notions are often used to explain the elusive history of Bigfoot—they are not normal, physical entities. And if Native American traditions are correct, these abilities are expected of a mystical *skinwalker*, a spirit able to take any form and to behave in ghostly ways. If Bigfoot creatures are shapeshifters, we may as well hang up our cameras, sit back, and wait for them to contact us. But while we’re waiting, let’s look at some other media stars…

The Monsters on Maple Street

Small-town America cherishes its neighborhood monsters. Some are whimsical; others are quite menacing. The mere possibility of their existence is influencing major life decisions. Whether a story is based on fact or fiction, once the tale grows legs (or paws) of its own there’s no stopping it.

In 1893, surveyor Eugene Shepard claimed to have captured a strange beast known as a *hodag* near Rhinelander, Wisconsin. Shepard displayed the animal at the first Oneida County Fair shortly after. The public was mesmerized by the “fearsome critter” with the head of a frog bearing huge teeth, the spiny back of a dinosaur, and an ominous spiked tail. Not until a group of Smithsonian scientists came to visit did Shepard admit the hoax. The taxidermic chimera of teeth, skins, horns, and clawed legs (operated by wires) had fooled almost everyone.

Yet still, the hodag accounts live on. The Rhinelander school mascot is a hodag. Statues abound from one end of town to the other. Thousands come to pay respects to the beast at annual festivals and the local hodag museum. Many level-headed citizens claim to have seen living hodags over the past century, crawling through its natural habitat, the bogs of Oneida

County. Some of these are even bigger and fiercer than Shepard's beast, supposedly attacking and devouring the unwary. Was Shepard's hoax based on a real animal? Will a living specimen ever be found? Two scenarios are possible: If a hodag is discovered, the town will continue the rich heritage of the hodag tradition. Or if it's never found—the town will continue the rich heritage of the hodag tradition.

Before the hodag began its prowlings, another monster was haunting the pine barrens of southern New Jersey. In the 18th century, a woman remembered only as Mother Leeds was struggling with twelve hungry kids and an alcoholic husband. Upon learning of her thirteen pregnancy, she cursed her unborn and thus was born the *Jersey Devil*. The deformed offspring developed a horse-like body, horned head, and bat wings. Eventually, it found its purpose, roaming the region and wreaking havoc on the population. Thousands of eye-witness accounts of the Jersey Devil have been recorded since then.

Shapeshifters and Skinwalkers

Millennia before Lon Chaney Jr. perfected the art, Enkidu of *The Epic of Gilgamesh* was able to appear as a ferocious wolf-like creature. Shapeshifters make excellent villains. How can you deal with a threat that may look like your mailman? Modern morphing technology in videos makes the phenomenon look believable yet imagine the anatomical problems. Werewolf movies bother me. Not from terror, but from wondering where the fur goes when he turns back into a man. I know it doesn't fall off. Does it suck back into his follicles? Does it magically fade away?

I think the fur thing pretty much establishes that shapeshifting (*therianthropy*) is a supernatural event with no comparable physical counterpart.[147] Shapeshifters in mythology

---

[147] No, not body growth or metamorphosis.

and fairy tales usually change back and forth into animals or other people. Also, witches and magicians can change a regular human into an object or animal. (How many frogs have been kissed in hopeful expectation?)

Wolves seem to be a good choice for a new shape. If I had my choice, it certainly beats a swan or frog. Medieval European myths eventually morphed into wolves having more of their humanoid features still intact. Peasants feared these beasts whom they believed to be demonic or servants of witches. To make matters worse, you never knew if your neighbor was a werewolf (*lycanthrope*). Some have suggested that the legends are based on a medical or psychological condition, or perhaps just a really stressful day. Haven't we all wanted to howl at the moon at some point?

Therianthropic concepts are common in many Native American traditions. The name "skinwalker" is derived from the Navajo belief in entities with the ability to appear in different forms, usually animal or animal-human composites. These shapeshifters are generally malevolent and greatly feared by some. Navajo shamans avoid using the name *yeenaldlooshii* which translates as "because of this it walks on four feet."

The Bray Road beast of Wisconsin is a wolf legend that refuses to die. Residents around Elkhorn's Bray Road have been reporting the werewolf-like creatures since the 1930s. A flurry of new sightings in the 80s and 90s brought worldwide attention.[148] Unlike the reclusive primate cryptids, this aggressive beast is more of a canine, yet still said to run bipedally like a human. The public was quick to note the similarity to the werewolves of movie fame. Alleged to be a relative of the Michigan Dogman, this 7-foot creature has a propensity for chasing cars (duh!). Explanations include bears, deformed bears, mangey bears, or hoaxers. As with similar

[148] Linda S. Godfrey, *The Beast of Bray Road: Tailing Wisconsin's Werewolf*, (Black Earth, WI: Prairie Oak Press, 2003).

phenomena, eyewitnesses "know what they saw!" and many are said to be quite reputable citizens with nothing to gain by deception.[149]

More Humanoid Haints

It does seem that some of the most disturbing cryptid reports are those involving creatures with human-like forms such as the Bray Road beast. Giant reptiles and fierce four-legged predators are certainly alarming, but beasts with a semi-human form are nightmarish. Perhaps these creatures represent archetypes of what we fear we ourselves might become under some extreme circumstance. Or do we have a hard-wired abhorrence of humanoids who don't appropriately bear God's image.

Another humanoid beast has more "goatly" traits and is said to lurk under the Pope Lick Creek train trestle near Fisherville, Kentucky. The Pope Lick monster, a beast with a terrifying, horned goat/sheep head and manly body allegedly lures unsuspecting teens hypnotically to the train tracks to kill them. Several young people have, indeed, died over the past forty years, either falling from the trestle or being hit by trains. Explanations for Sheepsquatch include a vengeful circus freaks, human-animal hybrids, and visitations by Satan himself. [150]

In the Ozark Mountains of Arkansas and surrounding states, residents report a feline cryptid dubbed the Ozark howler. Cat monsters seem a bit more believable than some others since in a face-to-face with a mountain lion, "monster"

---

[149] As with other large cryptids, I believe we can dismiss most hoaxers in costume with what I call the "just plain stupid principle." That's what it would take to put on fur and run through the woods in much of the country.

[150] Reminiscent of Well's *The Island of Dr. Moreau*, there have been suggestions that modern science (mad science?) might be responsible for some of these more bizarre beasts. In a coming chapter, we'll look at the possibility of human-animal hybrids. A number of paranormal cryptid reports allege secretive intervention by law enforcement or the military in containing or covering up information. In my opinion, this would seem to be more extraordinary than some paranormal explanations.

might be a good description. The howler, however, is alleged to be super-sized and somewhat humanoid, often running on two legs. Once again, reliable witnesses swear by their observations. On the skeptic's side, if I were being chased by a cougar, I might tend to overestimate the size and wouldn't likely get a good look at its posture.

Not every cryptid legend remains a mystery. The Loveland frogman is a good case study in cryptid lore. In the early 1970s, people around Loveland, Ohio reported sightings of a 4-foot frog creature. The beast could supposedly stand on two legs to escape when confronted. After decades of storytelling, apparently someone looked at the police report from that time. An officer had shot the animal and carried it away in the trunk of his car. He identified it as a 4-foot iguana (likely someone's lost pet) missing its tail. On the side of the folklorists, there really was a strange "monster" behind the myth.

One of the most famous paranormal creatures of all times harassed the town of Point Pleasant, West Virginia. The first sighting of the Mothman was by several men in 1966 who were digging a new grave. A few days later more people in the town began reporting a man-like creature with at least a ten-foot wingspan. Many witnesses commented on the beast's red, glowing eyes.

Most of the encounters were near an eerie, abandoned WWII munitions factory, considered an appropriate hideout for a devilish flying monster. Sightings continued for several months. Some skeptics attributed the phenomenon to a rare crane that had been observed in the area, or to secret night-time paratrooper operations. Paranormal investigator John Keel interviewed the townsfolk and found that some were experiencing random precognitions and visions. There were also reports of mysterious Men in Black (associated with the simultaneous UFO events) who threatened citizens, warning them to stop discussing any paranormal phenomena. The culmination of the Point Pleasant Mothman epic occurred on December 15, 1967. Forty-six people perished when the town's Silver Bridge collapsed. Many tied the disaster to the visions and Mothman. Subsequently, similar Mothman stories have occurred in other places around the world.

So how far can we push the borders of biological cryptozoology? Will our minds wrap around a cryptid such as a Jersey Devil or other more eccentric paranormal beings? Non-primate humanoids create problems. Paranormal Bigfoot at least had a physical, biological counterpart to consider. We know large primates can exist because they *do* exist. Our questions can be narrowed to whether they exist *here* and *now*.

However, *there are no real-world, biological counterparts to werewolves, sheepsquatches, mothpeople, or bipedal cats*. When we propose such creatures, they aren't simply on a continuum with biological cryptids—*they are something else*. Some of the creepier cryptid humanoids seem to move completely away from animal-like to just plain demonic (even by secular standards). Ghoulish *wendigos* and *rakes* are more the stuff of nightmares and Halloween than nature.

Then are we justified in suggesting that the Ephesians Hypothesis might be applied to some of these more esoteric cryptids? Many cryptozoologists, tribal philosophers, and media producers have already taken that quantum leap into another realm for us. And certainly, we must consider the Pickle Principle as well. A significant majority of paranormal cryptid incidents are likely misidentifications or hoaxes, as with other ghostly and demonic phenomena. But now, having connected cryptids, ghosts, and demons, would it be any more difficult or unexpected for a fallen spirit to deceive by means of a strange cryptid than by means of any other ghostly apparition?

Once again, without seeing demons behind every broken twig in the swamp, I believe the biblical worldview allows for fallen spirits that can and will use any deception within their power to lead the unwary astray. And as with other demonic phenomena, the more prone the "victim" is to believe in the esoteric supernatural, the more effective and deliberate the deception becomes. This could easily apply to the more "surreal" incidents in which physical laws are alleged to be bypassed. As with ghost phenomena, close-up cryptid experiencers often relate unique feelings in addition to the sensory data. Obviously, fear and alarm would trigger some hefty emotions, but many victims tell of a more "psychic" contact in which the creature is spiritually communicating or controlling.

Within the totality of cryptid reports, we would expect a broad range of observations and interpretations. It's one thing to believe that a stray black jaguar might wander up into Arkansas, yet quite another to seine the Mississippi in hopes of finding a mermaid. Wherever the cryptozoological consensus goes, we certainly should maintain a clear line between the possible existence of flesh-and-bone, biological cryptids and those that seemingly dip into the paranormal. I believe the common-sense application of the principles we've discussed establishes a good boundary for believability.

Merfolk and fairies would probably require a hefty sprinkling of pixie-dust to produce specimens, as would werewolves and goatmen. We'll see later that many attempt to connect cryptids to UFO phenomena. Is this link due to overactive imaginations or something more supernatural? If Bigfoot turns out to be an extraterrestrial, that would sort of make him a Wookiee, wouldn't it?

12
UFOs: The Lights' Side

"A UFO is the reported sighting of an object or light seen in the sky or on land, whose appearance, trajectory, actions, motions, lights, and colors do not have a logical, conventional, or natural explanation, and which cannot be explained, not only by the original witness, but by scientists or technical experts who try to make a common sense identification after examining the evidence."
Center for UFO Studies [151]

"There's a Starman waiting in the sky. He'd like to come and meet us. But he thinks he'd blow our minds." David Bowie [152]

Unidentified flying objects are just that—*unidentified flying objects*. And the truth is, as with other paranormal events, most UFOs are eventually *identified*. But let's be realistic. When the term UFO is used, our minds go directly to little green men. The idea that these unusual aerial phenomena carry visitors from outer space is known as the *extraterrestrial*

[151] Center for UFO Studies, www.cufos.org
[152] David Bowie, *Starman*, RCA Records.

hypothesis (ETH). The ETH has truly become *the* mythology of the modern era.[153]

Our earliest writings are rich with tales of mysterious strangers from faraway places. Even very young children need little encouragement to dream of fairylands with fanciful, unearthly creatures. We eagerly imagine places found only through rabbit holes or shrouded in clouds. It's not surprising that as we finally conquered the deepest, darkest locations on earth, our imaginations shifted upward to the unknown expanses of space.

A Very Brief History of UFOs

The modern-era UFO fascination took flight about seventy years ago. On June 24, 1947, pilot Kenneth Arnold observed nine unidentified objects flying in formation near Mount Rainier, Washington. In subsequent interviews, Arnold described the objects as "flat like a pie plate" with the back cut away as in a half-moon. He said their motion was like a "saucer skimming across water." With that, the term "flying saucer" was born, and the ETH reached escape velocity.

A few days later, there were several reports of strange objects flying around Roswell, New Mexico. On July 4, 1947, a farmer near Roswell reported wreckage from what appeared to be a crash strewn over his property. The site was investigated by officials from nearby Walker Air Force Base (then Roswell Army Base). At this point, the story branches into a myriad of scenarios. The military's stance, through several revisions, is that the event was the crash of a top-secret package of radio equipment carried by a balloon. Other accounts claim that alien

[153] For a number of reasons, the term *UFO* is often replaced with the more benign moniker *UAP* (unidentified aerial phenomena). The UAP designation is preferred by the government, military, and academia since it doesn't bring to mind flying saucers and extraterrestrials.

bodies (perhaps living) were recovered. Some believe this evidence is hidden at the infamous Area 51.

Needless to say, whatever happened at Roswell has surpassed legend status. It has become the quintessential UFO story, inspiring believers and debunkers worldwide. Today there are no surviving witnesses, and the event has generated so much hearsay and misinformation, on both sides of the argument, that the truth about what happened will likely never be known. Whatever did happen, the ensuing saucer-mania now has a life of its own, and a significant number of people are convinced of a cover-up.

But the events of 1947 were just the beginning of the UFO saga. In the chill of the Cold War, unidentified flying objects were seen as a threat—by both sides. One reason was that both the U.S. and Soviets feared retaliation by the other side if a UFO was mistaken as a military weapon. Declassified documents show that JFK discussed the issue with the Soviets just weeks before his death. Undoubtedly, much of the UFO secrecy within governments is still caused by the edginess of who-knows-what-and-how-much.

Perhaps as notorious as the UFOs themselves is the enigmatic role of the federal government in dealing with them. Conspiracy theories abound as the Feds "neither confirm nor deny." The mystique is fueled in part by the mysterious *Men in Black* who allegedly appear whenever a UFO flap arises. Many see these figures as agents of the presumed *deep state* which spins and censors details of UFO encounters. Nevertheless, the official stance of the U.S. government has always been that UFOs are simply mistaken identifications or yet-to-be-proven mistaken identifications.[154]

Despite government claims, however, we know through freedom-of-information policies that mountains of classified information on UFOs exist (in other countries, as well). Private

[154] Alternatives offered by the ufology community are 1) The Feds are as confused as everyone else, 2) ETs have been confirmed, but secrecy is necessary, 3) UFOs are secret deep state weapons.

individuals and some government leaders are currently demanding what is called *disclosure,* or the declassification of all UFO files. According to several former government and military employees, FOI guidelines are presently being circumvented by outsourcing UFO investigations to civilian companies.[155]

The Research

The U.S. government and military have performed numerous official and unofficial UFO investigations. The most well-known is Project Blue Book, carried out by the Air Force from 1952-1970. The objective was to scientifically analyze the UFO phenomenon and determine its threat level. Astronomer J. Allen Hynek served as scientific consultant for the project. The "official" conclusion of Project Blue Book was that no threat existed, and UFO phenomena could not be attributed to extraterrestrials. Afterward, Hynek produced a rebuttal of the findings insisting that the investigation was inconclusive.

Since the 1960s, a private group, the Mutual UFO Network (MUFON), has been recording and researching UFO phenomena. One of the leading, non-governmental organizations devoted to UFO studies, MUFON has over 120,000 cases in its database. As the Pickle Principle predicts, MUFON classifies many of these reports as misidentifications and hoaxes. Over 30% of MUFON cases are explained by misidentified natural phenomena.

Also, most sightings are classified as "not enough data to determine." About 60% of the cases cannot be studied because the evidence is anecdotal. This doesn't necessarily mean that

[155] The resulting network is often referred to as the military-industrial complex. For more info on government UFO research, see the works of Luis Elizondo, Stanton Friedman, Nick Pope, or Chris Mellon.

the evidence is faulty, but that it leads to a dead-end. For example, a couple witnesses an object floating over their house. Their account might be detailed, and they might be reliable pillars-of-the-community, but there are simply no further leads to explore. The sighting cannot be fully resolved, dismissed, or reclassified.

But herein lies one of the complications in analyzing UFO reports: *That remaining 5%—the ones that cannot currently be explained naturally.* Astrophysicist Hugh Ross uses the term *RUFO* (residual UFOs) in discussing these unexplained cases.[156]

It would be easy to conclude that these few remaining accounts are simply the cusp of the "60%-not-enough-data" group. In examining the cases, however, that doesn't seem to be the case. MUFON and other researchers point to a qualitative difference between this small group and the ones that simply can't be explained due to lack of data.

That 5%

Five percent might sound insignificant until we remember the tremendous total number of reports. That 5% RUFO group becomes *thousands* of noteworthy incidents. I must admit that in my research of the RUFO cases, there is ample substance for further study. The vast majority aren't lacking in data, so they can't convincingly be dismissed as such. And as mentioned, it's the qualitative aspect that differentiates them. There are "sets" of characteristics among these cases making them unique. Commonalities begin to emerge when the data are examined comprehensively…

---

[156] Hugh Ross, Kenneth Samples, and Mark Clark, *Lights in the Sky and Little Green Men, A Rational Christian Look at UFOs and Extraterrestrials*, (Reasons to Believe, 2002).

1. Multiple evidence types. Most sightings start with exactly that—a sighting. An unusual object or light is observed. But often there is more than just visual evidence. Sometimes there is radar confirmation of a flying object. Tracks, disturbed surroundings, burn marks, or residual radiation have been associated with some landing sites. Also, observers or those nearby may experience electrical anomalies in devices, sounds, and physical ailments. More recently, FLIR and other infrared technologies have revealed even more data.

2. Multiple, consistent observations. A group of supporting eyewitnesses is a dream-come-true for a trial attorney. How many court cases are won on consistent testimonies from several witnesses? I tend to pay attention when several people agree regarding a UFO event. No, I'm not so gullible as to assume groups can't be wrong or deceptive. But like any good juror, I believe the more observers, the less chance of error, fraud, and conspiracy. And allow me to make perhaps an obvious observation: Some witnesses are more reliable than others. All other factors being equal, UFO researchers tend to invest more "trust" capital in professionals (law enforcers, military, pilots, scientists). Honest investigators also favor information from those who have more to lose than to gain from reporting an incident.

When an event does have multiple witnesses, invariably the murmuring begins— "mass hysteria!" Whatever the popular belief, true mass hysteria isn't what most people are thinking when they use the term.

[157] The phenomenon can't logically be applied to the typical multi-observer UFO event, or for that matter, most other

---

[157] "Mass hysteria is a phrase that is used so often and so imprecisely to refer to anything from giving in to fashion fads to participating in riots and raves that it has become something of a fluid concept, synonymous with anything with a negative connotation that involves the participation of a large group of people." Maria Cohut,

paranormal phenomena. Moreover, I don't put a lot of stock in mass conspiracies. It's difficult for two people to keep a secret for very long, let alone dozens.

3. Complex actions. To understand this aspect, imagine releasing a fluorescent helium balloon (I could make one cheaply and quickly). It might look like a UFO to an unsuspecting observer, but the random floating motion wouldn't likely convince most people. Let's take it up a notch. Fasten the balloon to a toy drone. Now I can make more complicated, and perhaps more convincing moves. The behavior seems more deliberate. But the drone rig is more expensive and labor-intensive. *With complexity and added resources comes a lower chance that I would actually do this.*

If we extrapolate this to an object that darts between trees at high speed, splits into five separate objects that fly off in different directions and can make right-angle turns at breakneck velocities—well, I would like to play with that myself! And note, it's not that these features are impossible, but greater complexity tends to imply lower chances that someone could manage the hoax. This is especially true when some UFOs seem to maneuver in ways that defy the laws of physics or at least are beyond any technology known publicly.[158] Are there actually individuals or governments with the extreme resources to support the phenomena? If so, do they have sufficient motivation to make the investment or take the risk? I believe a synergy develops when all these characteristics are combined in the same event. As in any interpretation of research, the conclusions are more robust with more diverse sources of evidence.

"Mass Hysteria: An Epidemic of the Mind?" www.medicalnewstoday.com, July 27, 2018. Also, a common knee-jerk response by skeptics is *folie a deux*, or some other shared psychotic disorder. At some point even skeptics must admit the alternatives are far less likely than an *observationally accurate* event.

[158] I'm sure there are and have been talented individuals who would love to pass off a really good UFO hoax. The sheer number of successful "hoaxes" then becomes problematic. If some government has this advanced technology, why would they play around with it in such frivolous ways? Again, complexity doesn't *prove* anything, but it certainly tips the scale.

Case Studies

The facts of the Roswell account have been broadcast, headlined, denied, rediscovered, dramatized, rewritten, disclosed, and reborn to the point that probably no one knows the truth. But despite its renown, I don't believe Roswell has ever had the vigor of other lesser-known cases. In this section, we'll look at some case studies of the 5% scenarios. There are many extremely interesting enigmas in this group, and the evidence seems credible. I've attempted to summarize four that illustrate the qualitative characteristics mentioned previously.

1. The Rendlesham Forest Incident. Often called "the British Roswell," the events at Rendlesham Forest are perhaps the best-documented thus far.[159] These incidents took place in Rendlesham Forest, Suffolk, England in 1980. The six-square-mile forest lies between two air force bases, RAF Bentwaters and RAF Woodbridge, both being operated by the U.S. Air Force at the time.

On 26 December, 3 a.m., at the east gate of the RAF Woodbridge, base policemen John Burroughs and Bud Steffens responded to a strange red and blue light descending into the adjacent forest. The initial assumption was that a civilian aircraft had crashed since there were no military operations that night. The two drove toward the scene and noticed an additional white light moving through the forest in a manner uncharacteristic of an aircraft. At this time, they decided to inform the base. Burroughs and Steffens, not wanting to use their unsecured radios, drove back to the gate and used the landline phone. The message was relayed to flight chief Staff Sargent Jim Penniston who along with A1C Edward Cabansag, joined the two at the gate.

---

[159] Nick Pope, *Encounter in Rendlesham Forest*, (New York: Thomas Dunn Books, 2014).

Upon hearing Steffen's comment "It didn't crash. It landed," Penniston radioed the Master Sergeant J.D. Chandler who contacted other surrounding bases and Heathrow Airport. Radar had indeed detected a "contact whose identity is unknown" fifteen minutes prior to the incident and disappearing near Woodbridge. [160]

Penniston, Burroughs, Cabansag, and Chandler who had now arrived drove back into the forest toward the lights to resume the investigation. As they approached the lights, the men reported a static charge on their bodies and the loss of radio communication. At that point, Chandler and Cabansag returned to the vehicle to maintain radio contact with the base.

As Penniston and Burroughs came closer, they reported that the lights seemed to "explode" silently into a mass of very bright light. After the explosion, a column of light remained over Burroughs as Penniston moved even nearer. He said the spherical clump of lights was actually a metallic craft of some sort. It was roughly triangular, about three meters across and three meters high. Blue lights were aligned along the perimeter with a larger white light on top. The craft was completely silent. Penniston moved directly up to the object. He noticed unusual symbols on one side, which he sketched in his notebook. He recounts the event…

> The skin of the craft was smooth to touch. Almost like running your hand over glass. Void of seams or imperfections, until I ran my hands over the symbols. The symbols were nothing like the rest of the craft, they were rough, like running my fingers over sandpaper.[161]

In a few minutes, the object began to rise, slowly and silently above the treetops, then shot away at a very great speed.

---

[160] Pope, *Encounter*, 4.
[161] Pope, *Encounters*, 7.

Subsequent debriefings turned up many anomalies. Burroughs said that he remembered little of the incident from the light explosion (when he seems to have been knocked unconscious) until the object began rising. Also, Chandler and Cabansag had lost contact with the two others and the lights during a forty-five-minute period. They were ready to engage a search party when Penniston and Burroughs returned. However, Penniston and Burroughs claimed to have only been in the forest for a short time. Later they discovered that both their watches were slow by forty-five minutes.

In the following hours, more investigations revealed three indentations in the frozen ground where the craft was observed, presumably landing-leg marks. There were many broken branches in the path where Penniston and Burroughs witnessed the object navigating. Several rolls of photos were made by different people, but all were found too "foggy" to develop by the base darkroom workers. Initially thought to be tampering, some think the film may have been tainted by significantly high radiation levels later measured at the site. At 5:00 am, the situation was turned over to Deputy Base Commander Charles Halt. Lieutenant Colonel Halt, described as a stickler for procedure, filed all of the appropriate paperwork regarding the event using the term "unexplained lights."

If not already verging on "X-Filesque," the account becomes even stranger. The next night, Halt and much of the base were at an awards banquet. Reminiscent of the movies, the shift commander located Halt and told him "It's back."

Halt quickly left the function, later explaining that his intent was to settle the UFO distractions and get operations back to normal. Over several minutes, he assembled a small team (Burroughs also came) and ordered sets of large gas-powered lighting arrays sent to the forest.[162] The team drove directly to the site of the original sighting without observing anything unusual. As they investigated the area once more, a light appeared in the forest ahead. Recorded voices (Halt often used a small pocket recorder) describe a yellow light with "pieces falling off." The extremely bright light had a darker center that seemed to blink like an eye. When they were almost underneath, a thin beam of light from the object hit the ground beside them. They watched as the beam seemed to probe the base property. Burroughs later described that as the event ended, the same explosion of light filled the area.

I try to keep a critical, yet open mind regarding these incidents, but after studying the eye-witness accounts and other available documentation, perhaps I'm most amazed by the skeptics' responses. The explanations offered seem more contrived than any of the witness testimonies, not addressing any of the well-documented events. From lighthouse beams to joking airmen, the conclusions trivialize a large group of military personnel, with "no dog in the fight."[163] The possibility of some hi-tech prototype would explain much, but that seems inconsistent with the technology of forty years ago. Some have also suggested that if a cover-up was involved, it could be related to the possibility of nuclear devices maintained at the bases. Penniston, one of the main players, isn't convinced of ETs, but still has no good explanations…

> My thoughts are simply that 99 percent of all so-called UFO sightings can be explained by people with a knowledgeable background or aerial training to reporting such things for exactly what they are… this

---

[162] The following events were recorded on Halt's hand-held audio device. He claims in a later interview that much of the recoding has disappeared. Eighteen minutes of his descriptions of the UFO may be found on several YouTube sites.

[163] For example, at the time of the incident, Penniston had served through Operations Desert Shield and Desert Storm, and was the senior security officer at the base, possessing some of the highest military clearances. He was trained in aircraft recognition and parts of his career still remain classified.

> leaves the remaining one percent. It's this percent I believe is the truly unknown—a conclusion I made after I left the forest that night. This is the very reason I am troubled by the events of December 26, 1980… I left the forest with the "One Percent Factor" raining all over me.[164]

2. The Hillsdale Sighting. On 21 March 1966, around 10:30 pm, 87 girls in the MacIntyre Dormitory at Hillsdale College (Hillsdale, Michigan) observed a bright, silvery-white object over the adjoining arboretum.[165] For over ten minutes, the object displayed pulsating red and green lights, moving from side-to-side slightly. When the football-shaped light moved closer to the dorm, one of the girls called the Hillsdale Civil Defense Director Buck Van Horn. In the next minutes, Van Horn and several dorm moms joined the girls in one of the rooms and continued to observe the light.

Two Hillsdale police officers who saw the object from a mile away drove to the campus arboretum and documented the event. They also reported that the radio in their car stopped functioning. Eventually, all watched the object split in half, the segments traveling in different directions. The lights were observed until disappearing into the distance.

In an interesting side note, this event was investigated by Project Blue Book's Dr. J. Allan Hynek. Hynek officially deemed the sighting "swamp gas."[166] Those who had witnessed the phenomenon were incensed, by the simplistic and obviously contrived analysis. The public outcry eventually resulted in

---

164 Pope, Encounters, 11.

165 Jack Butler, UFO: "Hillsdale Had Its Own Close Encounter," *The Collegian*, www.hillsdalecollegian.com, March 19, 2015.

166 Project Blue Book was one of a the most extensive government studies of UFOs. It was conducted by the US Air Force from 1952 to 1969. The goals were to 1) determine if UFOs were a national security threat, and 2) scientifically analyze UFO data. Under the direction of J. Allen Hynek, thousands of UFO reports were collected and analyzed.

state representative Gerald R. Ford's call for a congressional hearing on the Air Force's handling of Project Blue Book. The hearing led to a new study of UFO activity, the Condon Report. This was also a turning point for Dr. Hynek. When eventually free from the alleged coercion of the Air Force, Hynek became a staunch believer in UFO phenomena and founded the Center for UFO Studies (CUFOS).[167]

3. The Phoenix Lights. There are several accounts of mass-sightings of UFOs over a long time span. The Phoenix incident involved thousands of witnesses over a 300-mile flight path. At about 8:00 pm, 13 March 1997, a large, triangular object was reported hovering over Henderson, Nevada. Reports mention 5-10 large lights on the leading edge of the object. Minutes later as it entered Phoenix airspace, numerous reports were relayed to authorities. Most witnesses saw only the lights in the same triangular configuration without the solid shape. The wingspan of the object based on the light array was estimated to be at least a kilometer. Over 2-3 hours, the object flew near Tucson and eventually Sonora, Mexico.

The phenomenon was well-documented with photos and video. It remains a complicated case due to the tremendous number of witnesses and government personnel involved. Several official explanations have been offered, from flares to plane formations. All are strongly disputed by witnesses.

With thousands of observers, there is no shortage of data. A large number of these witnesses were interviewed and can be found on many different videos. The descriptions seem remarkably similar. Most deviations can be attributed to distance and other viewing factors. I've collected representative statements from these interviews…

---

[167] Hynek speaking of the philosophy of CUFOS: "The attitude we are taking in the Center for UFO Studies is that since we are going to have scientists involved, we'll push the physical approach as hard and far as we can—instrumentation, physical evidence, photographs, radar records. If we are finally forced by the evidence to go into the paranormal, then we will." J. Allen Hynek, "Fate," June 1976, in *The Mammoth Encyclopedia of Extraterrestrial Encounters*, p. 304.

- Five lights passing overhead at a very slow rate.
- Too slow to stay in the air if it was an aircraft.
- "V" formation of five lights with two lights trailing.
- Two lights docking and undocking from the other five.
- So big we couldn't see both edges.
- I would gauge the object to be several football fields.
- A mile across—maybe more.
- Like we could actually reach up and touch it, it was so low.
- It was so low, all we could see was the left wing.
- Quite obvious it was no conventional aircraft.
- Soft glow. Bright, but no glare.
- Orangish-amber light, so lustrous it was almost like the objects seemed to be made of light.
- Three yellowish lights, perfectly round and equidistant.

A boomerang-shaped craft is common in UFO reports worldwide. The Phoenix incident is similar to the Hudson Valley lights (1983-1986) and the Belgium UFO wave (1989-1990). The shape would be easy to dismiss as an experimental B-2-type aircraft were it not for its immensity and flight habits. *4. The Nimitz Encounter.* On 14 November 2004, The *USS Nimitz* Carrier Strike Group and the missile cruiser *USS Princeton* were conducting drills about 100 miles southwest of San Diego, prior to Persian Gulf deployment. At midafternoon, two F/A-18F Super Hornet fighter jets from the *Nimitz* were ordered to stop maneuvers and proceed to new coordinates. It was later acknowledged that the Princeton had been tracking "anomalous aerial vehicles" (AAVs, in lieu of using the loaded term "UFOs") in the area for the past several days.

The *Princeton*'s senior radar operator Kevin Day admitted that radar detected over a hundred AAVs in the week leading up to the event. He claims that the radar images at times were "like watching snow fall from the sky."[168] Initially, the radar team suspected electronic glitches, all systems were tested and approved. The objects first appeared at 80,000 feet, which is well above typical air traffic, then dropped very quickly to the ocean surface.

When the diverted jets reached the site, the lead pilot reported a disturbance in the water with waves crashing over what appeared to be a large object just under the surface.[169] Around this same time, sonar operators reported a submerged object moving at twice the speed of submarine capabilities. Within seconds, another object appeared about fifty feet over the water. It was described as a white, forty-foot oval shaped like a Tic Tac candy. There were no visible wings or rotors. One of the pursuing pilots recounts the object's unusual flight characteristics…

> And what I mean by "erratic" is that its changes in altitude, airspeed, and aspect were just unlike things that I've ever encountered before flying against other air targets. It was just behaving in ways that aren't physically normal… aircraft, whether they're manned or unmanned, still have to obey the laws of physics…The Tic Tac was not doing that.[170]

The Nimitz events were part of the data leading some military and government officials to become more interested in UFO phenomena. Vehicles that can move at hyper-speeds, appear and vanish in our airspace, fly without exterior

---

[168] Interview in *Unidentified: Inside America's UFO Investigation,* The History Channel, 2019.

[169] Lead pilot David Fravor was commander of the elite Black Aces, a Top Gun program graduate, and had more than sixteen years of flight experience.

[170] Matthew Phelan, "Navy Pilot Who Filmed the 'Tic Tac' UFO Speaks," New York Magazine, www.nymag.com, 19 Dec 2019.

mechanisms, and out-maneuver our best technology deserve attention. At least that's what Senator Harry Reid and others concluded…

> I personally don't know if there exist little green men other places, I kind of doubt that, but I do believe that the information we have indicates we should do a lot more study...[171]

At the urging of Reid, a secret investigation of *unexplained aerial phenomena* (UAP… again, a less controversial term for UFOs) received funding of 22 million dollars. The Advanced Aerospace Threat Identification Program (AATIP) investigated data on UAP from 2007 to 2012. It was headed by Luis Elizondo, of the Office of the Under Secretary of Defense for Intelligence. He was instrumental in the DoD's release of previously classified aerial videos made during the Nimitz incident. In a surprising turn, the Pentagon (which had denied any such program during its operation) admitted in 2017 that the AATIP had existed.[172] Shortly after, Elizondo resigned his post in protest of the government's refusal to make full disclosure of AATIP to the public. In an effort to continue the AATIP's work, Elizondo established the private organization *To the Stars Academy*. As for the government's current role in UAP research, DoD spokesman Christopher Sherwood stated…

---

[171] Owen Daugherty, "Harry Reid Urges Senators to Push for More UFO Research," www.thehill.com, January 10, 2019.

[172] I love my country, and I understand the need for some secrecy. I see no need for outright lying. If anyone doubts that government individuals or departments do, in fact, lie, we need look no further than the AATIP. Under the auspices of the Defense Intelligence Agency, the UFO research program ran for five years from 2007 to 2012 with a budget of 22 million. Its function was to discern and analyze the threat posed by UFOs. AATIP was initiated by Senator Harry Reid at the lobbying of Robert Bigelow of Bigelow Aeronautics. The government and military denied the project's existence throughout its operation.

> The department will continue to investigate, through normal procedures, reports of unidentified aircraft encountered by US military aviators in order to ensure defense of the homeland and protection against strategic surprise by our nation's adversaries.[173]

Clash of the Real-World Worldviews

One of the themes of this book is that *worldviews have consequences*. How we observe, frame, and interpret our world really does affect decision-making. For example, Luis Elizondo believes that one factor contributing to the government's slow response in addressing the UFO threat is that some influential government and military leaders operate from a Christian worldview. He maintains that several high-ranking individuals, because of their Christian beliefs, don't believe ETs exist. Furthermore, these leaders interpret unexplained UFO phenomena as supernatural (God or Satan). Elizondo's U.K. counterpart Nick Pope notes the situation…

> I was aware that Pentagon pushback on UFO research was in part due to the religious belief of some of those involved… It was an odd irony that UFO investigations were being hampered because some people's belief in God meant that they either didn't believe in the existence of extraterrestrial life or that they regarded UFOs and extraterrestrials as demonic.[174]

---

[173] Steven Greenstreet, "The Pentagon Finally Admits It Investigates UFOs," www.nypost.com, May 22, 2019.

[174] Jasper Hamill, "Pentagon and MoD Officials Feared UFOs Were Either 'Demonic' or Sent by God, Former Investigators Reveal," www.metro.co.uk, May 8, 2018.

A future chapter will deal with the credibility of this claim. For now, though, I simply want to point out the worldview implications. As Elizondo and others warn, if UFOs are real vehicles, controlled by ETs or other countries, the potential security threat they pose should be of enormous concern. We can easily see in this situation two different worldviews and how they affect real-life responses. Decisions *are being made* regarding UFOs, and they are derived from the worldviews of those who make them. Worldviews based in reality (the true nature of things) will produce correct decisions. *Worldviews have consequences...*

13
UFOs: The Inside

"In the firm belief that the American public deserves a better explanation than that thus far given by the Air Force, I strongly recommend that there be a committee investigation of the UFO phenomena."
Gerald R. Ford, 1966.

"It got closer and closer and right above the pine trees it stopped and then it began to change colors from blue, to red, to white. Then it stayed there for a while. We were all aghast. We didn't know what it was. And then it just disappeared into the west. That was the end of it."
Jimmy Carter, on a 1969 incident.

"I occasionally think how quickly our differences worldwide would vanish if we were facing an alien threat from outside this world."
Ronald Reagan, 1987.

I realize the stigma attached to this topic. When serious thinkers or even presidents raise the issue, they may as well cover their heads with aluminum foil and buy a lifetime Comic-Con pass. As a science fiction enthusiast *and* a Christian *and* a scientist, I often ponder the intersection of these perspectives concerning extraterrestrial life. Yes, when I gaze into the starry night, the nerd part of me hopes that Ewoks are gazing back at me from a galaxy far, far away. But how do extraterrestrials (ETs, or more specifically ETI, extraterrestrial intelligence) fit with what God has revealed about himself and his universe? Has an unlimited God limited himself to a realm of a few billion humans on a lonely little planet?

Bottom line—Scripture doesn't provide specific information on this topic. We are assured there is at least one other "somewhere" beyond the one we now experience. The Bible gives us glimpses of another domain, where God and his angels now reside. It seems to be a world quite different from the one in which we presently live, and more amazing than anything science fiction can conjure.

The Theology of ETs

In his *Space Trilogy*, C. S. Lewis hypothesizes a solar system filled with *Hnau* (ETs). According to the storyline, these races never fell in sin as did the humans. Death, sin, and evil are dealt with differently than on fallen Thulcandra (Earth). Hnau are also able to intermingle with the angelic beings and have an unmarred relationship with God.

Extraterrestrial life does raise some interesting theological questions. Are humans the center of God's dealings? (Pre-Copernicans believed the earth was the center of the universe, but they were wrong.) If there are ETs, do they need redemption from sin? Did God also come incarnate as one of them, or do they have doctrines of a Savior who died for them on a faraway planet? Will they someday be our new mission field?[175] These

[175] I personally think Scripture indicates that humans are *it* for sentient beings in the physical cosmos. Note the description of Christ's work: "Nor did he enter heaven to

questions might sound farfetched, but the discovery of non-earthly beings would demand theological answers. *It would change worldviews.* Moreover, the mere *potential* of discovery is already altering worldviews. Pope Francis for one, is already planning for the possibility…

> If, for example, tomorrow an expedition of Martians came, and some of them came to us, here... Martians, right? Green, with that long nose and big ears, just like children paint them... And one says, "But I want to be baptized!" What would happen? When the Lord shows us the way, who are we to say, "No, Lord, It's not prudent! No, let's do it this way..."[176]

What would it be like if someday we discovered a race of ETs? Heiser suggests that perhaps ETs could also be created "in God's image" since the phrase may be translated "as God's imagers."[177] Could God have other imagers besides us?

Despite our speculations thus far, extraterrestrial biological beings are conspicuously absent in Scripture, yet we shouldn't build an argument from silence. The Bible quite often portrays humans as a singular component in God's creation. God's grace, mercy, and love seem to be uniquely poured out on humans…

> For surely it is not angels that he helps, but he helps the offspring of Abraham. Therefore, he had to be

---

offer himself again and again, the way the high priest enters the Most Holy Place every year with blood that is not his own. Otherwise Christ would have had to suffer many times since the creation of the world. But he has appeared once for all at the culmination of the ages to do away with sin by the sacrifice of himself. Just as people are destined to die once, and after that to face judgment, so Christ was sacrificed once to take away the sins of many" (Hebrews 9: 25-28). Nevertheless, I keep a slightly-open mind.

[176] Abby Ohlheiser, "Pope Francis Says He Would Definitely Baptize Aliens If They Asked Him To," www.theatlantic.com, 12 May 2014.

[177] Michael Heiser, *The Unseen Realm: Recovering the Supernatural Worldview of the Bible,* (Bellingham, WA: Lexham Press, 2015).

> made like his brothers in every respect, so that he might become a merciful and faithful high priest in the service of God, to make propitiation for the sins of the people (Hebrews 2:16-17).

This passage is limited to a comparison with angels. but the implication seems broader.

I frequently hear the argument: "If Earth is the only planet with conscious beings, then why is the universe so enormous?" This question doesn't take into account God's revealed nature. Our Creator is extravagant. His resources are boundless. He often goes above and beyond merely adequate to show his glory and his love for his creatures. Flowers grow in fields unseen by a single eye. Amazing fish are doing amazing things miles below the ocean's surface in total darkness. Besides, it's no more difficult for God to make a universe than a flower! Ron Rhodes writes that God's purpose behind the vast universe of planets is…

> …not to inhabit them with multiple life-forms but to serve as a testimony to his power and glory. It is not necessary to argue that simply because many planets exist in the universe on which life *could* exist, life *does* therefore exist on those planets.[178]

The Psalmist certainly understood the purpose of God's loving "overindulgence" in his creation: "The heavens declare the glory of God."

Personally, the adventurer in me would like to meet some aliens out there somewhere. Who knows—in the vastness of Eternity someday, will God choose to populate a few galaxies with other beings? We are his image-bearers—showcases for

[178] Ron Rhodes, *Alien Obsession: What Lies Behind Abductions, Sightings, and the Attraction to the Paranormal*, (Eugene, OR: Harvest House, 1998), 40.

his mercy and grace. Will other life forms in the eternal realm be used to showcase other attributes of the Godhead? I can only imagine. Whatever the case, we can rest assured that any beings we might encounter, now or later, will be God's creation and will be as much under his sovereignty as are we.

The Rime of the Ancient Aliens

The belief that advanced ETs have been visiting the earth for thousands of years results in some bizarre, and somewhat amusing conclusions. In 1968, Swiss hotel worker Erich von Däniken released *The Chariots of the Gods*. His premise was that all the "deities" in human history (including the God of the Bible) were actually extraterrestrials. According to von Däniken, these *ancient astronauts* were responsible for all the unexplained and paranormal artifacts of the past, such as Stonehenge, the Great Pyramids, Easter Island, the Serpent Mounds, the Nazca Lines, and the list goes on into his subsequent books.

Biblical miracles were also part of von Däniken's theory. Ezekiel's wheel is interpreted as a spacecraft. Elijah's fire from heaven was a phaser set on "kill." The parting of the Red Sea, manna from heaven, healings, prophecies, and even the creation of human civilization are all seen as extraterrestrial acts. This ideology demonstrates how errors in primary worldview components can drastically affect the interpretation of events. Nonetheless, the musings of von Däniken have huge appeal (70 million books sold), especially to anyone doubting or rejecting biblical explanations. What better way to quench the craving for some sort of transcendence without all the accountability guilt!

More recently, the ancient astronaut theory was taken to new heights by Zecharia Sitchin. A popular voice on many radio and TV programs, Sitchin used ancient texts to back up his claims. Perhaps the most significant boost for these ideas came from the History Channel's *Ancient Aliens* series. The 179 episodes (so far) have preached to millions this gospel of alien creators and saviors.

The idea that extraterrestrials "planted" life on earth is a popular sci-fi pitch. When serious scientists consider the idea, they use the term *directed panspermia*—the theory that life originated *out there* and eventually got *here*. Oddly, outspoken atheist Richard Dawkins supports this concept that super-intelligent aliens could be responsible for humanity, yet he flatly denies divine creation.[179] Unfortunately, the naturalist's problem with designed, purposeful creatures is simply pushed back one rung on the philosophical ladder.

Obviously, many problems are raised by ancient astronaut stories. Both von Däniken and Sitchin have rejected any notion of the biblical God. Many works have been exclusively devoted to debunking their ideas.[180] Archaeologists, historians, and academics in general point to the contradictions, misrepresentations, and lack of any scientific rigor in their work. Nevertheless, many with susceptible worldviews are being lured into UFO religions based on superficial analyses of the material.

The Science of ETs

Some Christians might worry that the reality of aliens could potentially affect our theology, but the lack of any such discovery is already causing sleepless nights for naturalists. If biological evolution is so forthright and apparent, shouldn't space be teeming with life? As mentioned earlier, the universe is an unbelievably huge place that likely has a staggering

---

[179] Jeff Miller, "Directed Panspermia and Little, Green (Non-Existent) Men from Outer Space," www.apologeticspress.org

[180] See the articles by biblical languages scholar Michael Heiser.

number of planets suitable for some type of life. If sentient life does exist elsewhere, are we coincidentally the most advanced civilization in our galaxy? Are our neighbors still in their Middle Ages? Assuming a large number of suitable planets and the principle of mediocrity, shouldn't there be a *lot* of ETs out there that are more advanced than humans?[181] If so, shouldn't we at least be detecting their electromagnetic signals?

In 1961, astrophysicist Frank Drake proposed an equation for estimating the probability of inhabited planets. Although its variables are difficult to determine, the Drake Equation did stimulate research in the area. Drake, Carl Sagan, and others have invested millions in the Search for Extraterrestrial Intelligence (SETI) project. Radio telescopes and vast antenna arrays have transmitted and listened intently for decades. A special message from humanity was directed into deep space in 1974. The Arecibo Message consisted of recorded icons of earthly life, hopefully, understandable to alien listeners. We're still awaiting a reply. If anyone is out there, they certainly are quiet. The embarrassing implication (to naturalists) is that we truly are alone in the physical universe. Maybe the evolution of life isn't as likely as theorized. Maybe humans *are* quite special.

It's interesting (and a little amusing) that science fiction writers are compelled to address this trade secret of cosmology. How can you have stories with alien life forms so diverse and believable when experience indicates that ETs are either very rare or non-existent? One way is to set the scene "a long time ago in a galaxy far, far away." Out of sight, out of mind. Another is to do what creator Gene Roddenberry did with *Star Trek*. The Prime Directive clearly mandates that advanced civilizations may not interact or even be detected by the less advanced. Do we simply have a lot of growing to do before ET phones us?

---

[181] This principle assumes that all things being equal, we should consider a random sample (in this case Earth) to be one of many "in the middle of the curve."

Lacking any solid evidence of sentient life in the cosmos, why should Christian apologetics devote time and energy to the question of ETs? I reiterate the rationale for examining the paranormal in general: Our beliefs affect our worldview. And of all paranormal phenomena, UFOs and ETs are perhaps the most endemic in our culture. Two decades ago, Kenneth Nealson, of the National Academy of Sciences was already asserting that ETs were "no longer a fringe type of thing."[182] And as previously noted, significant numbers in our population are already fully acclimated to the "fact" of extraterrestrial life, even without solid evidence.

As a baby-boomer, I can attest that TV and other media, in general, have nourished me with a steady diet of UFOs and aliens as far back as I can remember. For many, this modern mythology significantly blurs the boundaries of reality. A national survey in 2019 found that a third of Americans believe UFOs are extraterrestrial visitations and 16% claim to have seen a UFO or ET. Over 65% believe the government is covering up evidence related to UFOs![183] *Whether extraterrestrial life exists or not, its influence on our culture's collective worldview is significant.*

The large number of unexplained, yet convincing UFO accounts lead to only three viable realities, each with significant worldview implications…

- A group or a government with extensive technology and resources is implementing an extreme and deceptive agenda.
- Extraterrestrial life does exist, and we anticipate huge changes in our view of reality.

[182] Jim Wilson, Popular Mechanics, July 1999.
[183] Lydia Saad, "Americans Skeptical of UFOs, But Say Government Knows More," www.news.gallup.com, September 6, 2019.

- Paranormal or supernatural realms exist, and there are beings among us with a covert agenda.

A Grand Scheme?

Are UFOs and ETs just a clever ruse associated with some ulterior motive? Conspiracy theories abound in ufology. Publicly, the US, Soviet, and Chinese governments have all accused each other of being behind the phenomenon. It's a fact that governments hide the truth quite often, and there's no end to contradictory statements. A few earlier UFO sightings are now known to have been cold war projects (American and Russian). For years before the unveiling of the B-2 stealth bomber, people who reported its secret flights were scolded and muzzled by officials. The military staunchly denied the existence of any such craft. Afterward, all lies were justified as "routine security measures."

Does our government have new, secret technology? I hope so. And I think I would like for it to be somewhat secretive until we need it. Or is the secret that they know about and have communicated with ETs? Even more bizarre, has the government hijacked an alien vehicle? (This would make a good plot for a TV series!) I'm not dogmatic about it, but I tend to doubt these possibilities. The technology associated with what UFOs are observed to do is orders of magnitude beyond ours. Some UFO behavior even defies universal physical laws. The crafts have been observed to travel at speeds in the tens of thousands of miles per hour yet are able to turn right angles without slowing. At times they materialize or fade from sight, visually and on radar. Even at high speeds and near the ground, the objects produce no sonic booms or other sounds. Many observers report that the objects change shape, break apart, merge, pass through solid objects, and perform other unusual acrobatics. Some suggest projected or holographic sources, but the scale is still far beyond known abilities.

Considering the case studies in the previous chapter, each involved technology that was far ahead of the technology of the day. We might imagine experimental vehicles that are the next step up from our current known science, such as the early B-2 tests. But the "unearthly" machinery observed in these incidents is difficult to explain as "next generation." Some suggest that the Nimitz incident could be attributed to extraordinarily high-tech designs that employ superconductors, quantum gravity fields, inertial insulators, and a lot of other fancy-sounding science fiction gadgets. A popular notion is that someone might be "reverse engineering" alien technology to make such great strides in a short time. I suppose any of these are feasible at some level, but then we are faced with trying to explain a virtually super-human conspiracy. If that card is to be played, why not just go with the traditional UFO theory? And, we're still stuck with the anachronistic problem. If some government has been staging the phenomena, why would this advanced technology show up in the 1960s and remain unconfirmed and unchanged for over fifty years?

This points to a related puzzle. Perhaps the most compelling reason to doubt a widespread conspiracy is the span of UFO incidents. The phenomenon has outlived government "regimes" for over 70 years in almost every country. Again, proposing a conspiracy of such secrecy and effectiveness takes more gumption than simply believing in ETs. The governments with which I'm familiar have trouble managing healthcare and postal systems. They don't seem capable of keeping secrets through an election period, much less a half-century. It's easy to speculate about government UFOs, yet quite another to consider it pragmatically.

We could also ask why, if the government or military were able to pull this off, would they seem so disorganized in the actual responses to the phenomena? We've observed compartmentalization and maybe even petty competition among levels of government and military before, but not to the point that one branch would deceive and endanger individuals in another. (I'm hoping that only happens on TV.) There have been lower-level officials caught in cover-ups and disinformation, but paper trails quickly uncover the motives. So far, despite many efforts, no one has produced any verifiable, human scheme behind UFO activity. Once again, longevity, incompetence in secret-keeping, and context would seem to be inconsistent with a major conspiracy.

Does anyone have anything substantial to gain by a widespread infatuation with ETs (other than the Syfy Channel)? UFOs do seem to "want" to be known. But if they want us to notice, why be so elusive? If they're waiting for us to be "conditioned" in some way, we're way past that. Why not stage a first contact on the White House lawn tomorrow and get things moving? "We're ready!"

On the other hand, if their agenda requires secrecy, can't they do a better job than what we've seen over the past decades? I question the competency or the sincerity of a group whose agenda seems to be playing obscure games with random, usually uninfluential people. Despite all these improbabilities, UFO phenomena are real on some level. *Are we missing something?*

14
UFOs: The Other Side

"One theory which can no longer be taken very seriously is that UFOs are interstellar spaceships."
Arthur C. Clarke [184]

"I have come to support less and less the idea that UFOs are nuts-and-bolts spacecraft from other worlds."
J. Allen Hynek[185]

The extraterrestrial hypothesis is by far the most popular of the pop culture explanations for UFOs. Aliens are just plain cool—at least in stories. But as much as I would like to warp around the galaxy, seeking out new life and new civilizations, I must boldly go where I don't want to go in saying: *It probably won't happen...*

...in the physical realm anyway. I'm not an expert in physics, but my physicist friends have some discouraging words about interstellar space travel. Space is much bigger than words and numbers convey. The nearest star to our star (the sun) is 4.5 light-years away. (Just to enjoy the zeroes, that's 25,000,000,000,000 miles.) Let's say I have a spaceship that

---

[184] Arthur C. Clarke, "UFOS Explained," www.nytimes.com, 27 Jul 1975.
[185] Ian Ridpath, "Interview with J. Allen Hynek," *Nature*, October 4, 1975.

can go 50,000 mph. It would take me 500 million years to make a one-way trip to our nearest stellar neighbor.

Can we go faster? Maybe, but physicists say we can't surpass the speed of light.[186] And even if we could get *near* the speed of light, journeys to the rest of our galaxy would still take thousands of years. Many other factors will prevent interstellar travel beyond the single-digit light-year range, and even then, the trip would be perilous and very long. All of this to say that ETs face the same problems. The laws of physics are, indeed, *laws*, regardless of your planet of origin.

So, let's look at our paranormal options again:

1. Mistaken identification.
2. Directed human activity.
3. Observationally accurate event.

Let's also keep our premise that the majority of UFO cases are mistaken identifications or directed human activity. If not interstellar travelers then, do we still have a convincing explanation for the notorious "5%" that seems to be observationally accurate?

There have been several hypotheses put forth that sidestep the ETH, although they don't bypass the paranormal. Some have postulated that UFOs could be time travelers from our own future. Others speculate that ETs are as earthly as the rest of us, living deep in the ocean or in subterranean cities. But these options seem to pack the same excess baggage as the ETH.

---

[186] Gene Roddenberry solved the problem in *Star Trek* by having warp speed, the ability to travel multiples of light speed. An excellent idea, but it's imaginary. It wouldn't have made a very interesting series had the crew been limited to our solar system

The Interdimensional Hypothesis

If saucers aren't your cup of tea, the most popular *non-*ETH idea was popularized by Jacque Vallee in the 1970s…

> …are we dealing instead with a parallel universe… From that mysterious universe, are higher beings projecting objects that can materialize and dematerialize at will? Are UFOS "windows" rather than "objects?"[187]

Vallee's *interdimensional hypothesis* (IDH) provides a popular alternative to extraterrestrial visitation. Could we be observing vehicles from other space-time dimensions or rather the doorways to such? Some of those fuzzy orbs and blurred lights do appear too ephemeral to be solid objects.

Do advanced entities have the ability to navigate between worlds using dimensional windows? We've long envisioned worlds through looking glasses and over rainbows. The destinations are similar—a place unlike here. Far away, yet so near… And, other realms provide diverse narratives. The rules can be different in a different place. The old history doesn't determine the new present, as creators of superheroes know so well. The Marvel Universe obviously isn't bound by the limitations of our own.

Is the preoccupation with aliens preventing research in other areas?...

> In Europe UFOlogists have recognized these paraphysical characteristics for years and have made considerable progress by discarding the ET myth and

[187] Jacques Vallee, *Passport to Magonia: On UFOs, Folklore, and Parallel Worlds*, (Chicago: Contemporary Books, 1993).

> other aimless controversies that have bogged down their American counterparts...[188]

Before tumbling too deeply through this portal, let's consult the physicists again... Our human senses allow us to experience four dimensions—three spatial and one temporal (time). But multiple dimensions aren't inconsistent with modern physics and mathematics. String theorists, for example, propose at least a ten-dimension system in order to incorporate both quantum activity and gravity. Also, most cosmologists today accept an extradimensional cause for this universe. Whatever triggered the inflation of our space-time universe had to have been outside this 4-D universe.

The biblical worldview describes at least one additional dimension beyond our physical four. Moreover, the premise of this book is that there exists a dimension outside the one we perceive. Defining other worlds in science-ese doesn't negate the fact that when we speak of the heavenly realm we are speaking of another dimension—a real place that is beside/trans/beyond where we are now.[189]

Jesus himself provides a possible illustration of interdimensional movement if we aren't afraid to apply somewhat scientific terms to spiritual events. This passage describes one of his post-resurrection appearances...

> On the evening of that day, the first day of the week, the doors being locked where the disciples were for fear of the Jews, Jesus came and stood among them and said to them, "Peace be with you." (John 20:19).

The text indicates that Jesus apparently materialized abruptly in their secured hideout. Already in his glorified

---

[188] John Keel, *The Complete Guide to Mysterious Beings,* (CreateSpace Platform, rev. 2016), 154.

[189] Words certainly carry baggage. If you are uncomfortable thinking of parallel dimensions and hyperspace, just insert the phrase "heavenly realm" (or vice versa).

"spiritual" body, this was not simply a vision or other noncorporeal specter. Jesus's new body could be in the supernatural realm then suddenly pass into the natural. Later in a similar event, he even invites the inquisitive Thomas to touch his real, physical body: "Reach out your hand and put it into my side. Stop doubting and believe."[190]

So, as Vallee postulated, do other dimensions account for some of the unusual behavior observed with UFOs? Others have made this speculation in light of their speed, maneuverability, and ability to pop in and out of detection. Interdimensionality, as untestable as it is, would explain many unusual observations in a variety of paranormal events.

Keep in mind as I ramble, there isn't a shred of empirical evidence that other dimensions exist.[191] *If* they do, many scientists doubt that different dimensions can interact with each other physically, on a macro level anyway. (Bad news for *Stargate* and a gazillion other sci-fi episodes.) Others speculate that subatomic wormholes and quantum entanglement may hint at the possibility. What would the interface between two worlds look like? Would it be some sort of invisible energy vortex, only discerned by instruments? Perhaps the patriarch Jacob could have witnessed such a phenomenon in the form of a "stairway to heaven…"

> There was a stairway set up on the earth, and the top of it reached to heaven. And behold, the angels of God were ascending and descending on it! And behold, the LORD stood above it (Genesis 28:12-13).

Portals are a common feature in many fairy tales and paranormal accounts—a magical door connecting two realms.

---

[190] We also see in Scripture Enoch being miraculously transported to the heavenly realm (Genesis 5), and Peter suddenly whisked from Ethiopia to Azotus (Acts 8).
[191] Except the communicated evidence revealed in Scripture. This is the same evidence that allows us to know the supernatural in general.

In the *Stranger Things* series, a portal, or rift, allowed access to the Upside-Down world and its terrifying creatures. Theoretically, something can pass through a portal and suddenly exist in the other realm and not exist in the former. Or, if light can pass through, we could see into the other world like looking through a window. Maybe that's being described in Revelation 6 where "The sky rolled back like a scroll." I believe I'll experience a functioning portal someday when my spirit is suddenly "absent from the flesh, but present with the Lord."

Connections between worlds may be of the same nature as miracles. Like other biblical supernatural events, portals had a specific purpose at a specific time. It's serious business when God allows our realm to connect with his. Jacob's stairway was necessary to convince him of the spiritual side of reality. His observation of the interface was necessary for his understanding of God's plans for him and his progeny.

Considering Christ's enigmatic movements, his apostles needed a clear understanding that divine resources were not miles away in some faraway celestial city. Christ would not be distant! Their new heavenly portal would be as intimate as the Spirit himself living within them. As temples of the Holy Spirit, believers today have that same built-in corridor connecting us with the unseen "extradimensional" realm.

Experientially, dimensional portals seem to explain some odd UFO phenomena. If the UFO could be a manifestation of an object in a different realm, it wouldn't be limited by our physical parameters. It could appear to exceed speed limitations and bypass inertial and gravitational laws. An object could also seem to appear and reappear. Or perhaps as Vallee theorizes, the UFO itself may be the portal, moving and behaving apart from the laws of physical objects.

From a biblical worldview, however, we must also consider this in light of the Ephesians Hypothesis. Are fallen spirits allowed to manipulate and use such features as did the angels in Jacob's experience? I don't find any specific biblical examples of this, although the assumption is that demons can enter the physical realm, possibly in physical bodies. Also, we know that angels may enter our dimension with at least the appearance of physicality. If God does allow divine entities to navigate between realms in this way it doesn't seem that it would be inconsistent with Scripture. We'll explore this idea further in the following chapter.

15
UFOs: The Dark Side

"I thought that they were angels
But to my surprise,
They climbed aboard their starship
And headed for the skies..."
Styx [192]

"I was taken from my bed by two little beings who were two and a half, three feet tall. They came into my bedroom, took me through the living room... It was like we were floating, not walking. I had no control. They took me aboard this craft and gave me an examination... They took a sample from my leg. I have a scar on my thigh where they took a sample of tissue. They did something to my back and other things I'd rather not talk about." Shirley A. Coyne [193]

From his research files, J. Allen Hynek devised a rating system for UFO phenomena which was picked up by Hollywood in 1977. Stephen Spielberg's *Close Encounters of the Third Kind* included bits of the real lives of Hynek and his colleague Jacques Vallee. Hynek's UFO incident scale began with what he termed a "close encounter of the first kind" (CE-1), defined as an observation of a UFO up close (near enough

---

[192] Styx, *Come Sail Away.*
[193] Phil McCombs, "Secrets in the Saucer," www.washingtonpost.com, 29 Jun 1987.

to make out basic features). A close encounter of the second kind (CE-2) included some interaction with the environment such as a track, burn mark, radio static, or illness. According to Hynek's system, at a CE-3 level, an actual living entity associated with the UFO is observed.

Hynek's Extended Classification of UFO Close Encounters

1st kind – Observation of a UFO.
2nd kind – Observation of a UFO with physical evidence.
3rd kind – Observation of a living ET.
4th kind – Abduction by an ET.
5th kind – Direct verbal or mental communication by an ET.
6th kind – Death of human or animal caused by UFO.
7th kind – Creation of a human-alien hybrid.

Until now, our focus has mostly been on the nature of the UFO itself—that unidentified flying *object* shooting through the sky, or the strange light hovering in a forest. But a CE-3 is a boundary crossed. The ETH now has legs, and the event moves from a phenomenon with potential natural explanations to a narrowed notion that now must account for an intelligent, interactive source. When we examine CE-3 scenarios, there is also a shift in perspective which brings us much closer to the paranormal. Those who don't follow CE-3 incidents closely are often surprised to learn what the collective data reveal. One poll indicated that as many as 3 million Americans claim to have memories of being abducted or physically manipulated by what they perceived to be aliens.[194]

A sobering trend emerges among alleged extraterrestrial encounters. The ordeals are almost always dark, ethereal, and sinister. Experiences are no longer social events, instead, they're intimate and intrusive. Not only are the encounters disturbing simply because they are unfamiliar confrontations, but they also take on a transcendent, surreal, and confusing nature. Those who claim to encounter ETs don't walk away enthused or uplifted. The victims are certainly changed, but the experience is almost exclusively pathological. Vallee summarizes his investigation of these cases…

> The experience of a close encounter with a UFO is a shattering physical and mental ordeal… New types of behavior are conditioned, and new types of belief are promoted. [195]

The Hill Abductions

One of the earliest and most publicized abduction accounts involved Barney and Betty Hill in 1961. The two were driving home from vacation to their home in New Hampshire when Betty noticed a bright light that seemed to be following them. Barney stopped and got out of the car for a better look. As the object came closer, he could see that it was a large saucer-shaped craft with windows along the edge. Barney claims he could see odd, uniformed beings watching him through the windows. At that point, he ran back to the car and they hurried away as the saucer followed. They both told of strange beeping

---

[194] Budd Hopkins, David Jacobs, and Ron Westrum, "Unusual Personal Experiences: An Analysis of the Data from Three National Surveys Conducted by the Roper Organization," (Las Vegas: Bigelow Holding, 1992).

[195] Jacques Vallee, *Messengers of Deception: UFO Contacts and Cults*, (Berkeley: And Or Press, 1979), 9.

sounds coming from the trunk and the feeling of some type of energy passing through their bodies. The next thing the Hills remember was regaining consciousness about thirty miles down the road. Upon arriving at home, they realized the trip had taken two hours longer than expected.

In the following days, Betty began pouring over books about UFOs. After reading of the National Investigations Committee on Aerial Phenomena (NICAP), she contacted the group and filed a report of their encounter. Over the next years, Betty experienced vivid nightmares in which she and Barney were onboard a craft undergoing some type of medical examination. Betty underwent months of psychological therapy, wrestling with "new," horrifying memories. Much of her treatment involved hypnosis. During the sessions, she eventually pieced together more details of the couple's ordeal…

> During the examinations, the beings… plucked strands of their hair, took clippings of their nails and scraped their skin… Needles, connected to long wires, probed their heads, arms, legs and spines.[196]

Betty Hill's hypnotherapy uncovered many fragments of her encounter. She described grayish, large-head beings, and remembered talking at length with them. During the incident, she claims a long needle was inserted into her abdomen. One of the beings explained that it was a pregnancy test. In her stories, the aliens also had a sense of humor, laughing at Barney's removable dentures.

Another curious aspect was the star chart given to Betty by one of the creatures. Under hypnosis, she redrew a map of star positions that had been placed into her mind. After the publishing of the diagram, an amateur astronomer announced

---

[196] Linda Lacina, "How Betty and Barney Hill's Alien Abduction Story Defined the Genre," www.history.com, updated 15 Jan 2020.

that the map was from the perspective of a viewer in the star system Zeta Reticuli. Betty's "followers" still see this as proof of her encounter with extraterrestrials. Many astronomers, however, question the accuracy of the map.

Barney Hill died eight years after the event. Betty claimed to have been visited several times by the aliens until her death in 2004. A New Hampshire historical marker stands at the site of the alleged abductions.

The "Perfect Abduction"

The Hill case contained many elements that seem to be common in alleged abduction incidents since then. Skeptics claim that the Hills' highly publicized story may have influenced subsequent reports. In researching the material for this book, I reviewed over a hundred abduction cases selected from a variety of sources. With this base, I then followed Tyrrell's strategy for modeling a *typical* apparition. The "perfect abduction" description that follows is a mosaic of many abduction scenarios. Some of my categories are based on the observations of psychologist Joe Nyman who worked extensively with those claiming to be victims.[197] (For simplicity, in the remainder of this section I'll drop terms such as "claim to" and "allegedly" and assume the subject's point of view.)

1. The event starts late at night.
2. The victim is in his car or bedroom.
3. Immediately before the incident, the victim senses dread, general anxiety, or physical uneasiness.

[197] Nyman, Joe, "A Composite Encounter Model," In: Pritchard, Andrea & Pritchard, David E. & Mack, John E. & Kasey, Pam & Yapp, Claudia, *Alien Discussions: Proceedings of the Abduction Study Conference*, (Cambridge: North Cambridge Press), 83-85.

4. A dreamlike, ethereal feeling envelops. Voluntary muscle control is lost.
5. The victim is transported into a room or wakes up there. Usually if awake, he floats in a paralyzed state. Some recall passing through solid objects such as walls or ceilings.
6. The environment is foggy and surreal. Unusual lights are present. These tend to be bright until the abductee reaches the destination, then it becomes much dimmer.
7. The victim experiences laying on a surface and being "probed" with various metallic devices. Most attention is focused on the head and reproductive organs.
8. A "sermon" is given, often telepathically, in which the abductee is "enlightened" about a variety of vague philosophical topics.
9. The victim is returned to the place of abduction or nearby.
10. A loss of time is noted.
11. Subsequent abduction experiences are likely.

Those who have had up-close encounters with aliens are referred to as *experiencers,* or if manipulated in some way against their will, *abductees*. *Contactee* refers to a person who has communicated (perhaps telepathically) with an extraterrestrial. The descriptions are sometimes used interchangeably, but those who feel violated by ETs sometimes resent the neutrality of being considered a mere experiencer. It should be noted that quite often abductees and experiencers don't consciously recall their encounters. In many cases, the memories reside "deep in the subconscious" and can be recovered only through hypnosis or other treatments. Hypnotherapy isn't without its doubters. Some point to the problem of *false memories* that are known to arise during hypnosis. Experiencers can also be significantly influenced by the therapists' styles of questioning and other external suggestions.

Subjects with no conscious recollection of an alien contact may have dreams, flashbacks, and general paranoia which lead to suspicions and fears that an alien encounter could have occurred in the past (especially when hearing of similar symptoms in other abductees). It's also common for an experiencer to be a *repeater* or one who remembers several different encounters. Repeaters generally recall that their incidents started in childhood. On considering the characteristics of ET confrontations, skeptics often point to the similarities between experiencers and those with mental illness.

It's easy to see how a person with some form of mental illness could express the condition with the same perceptions. This in no way implies that all experiencers have a mental issue, but it's likely that some accounts may be explained as such. It should be noted that very many experiencers and abductees have no indications of mental illness whatsoever. Harvard psychologist John Mack who studied over 200 abductees concluded that psychological issues before his patients' abductions were not different from those of the general population. His controversial work with alleged abductees provided the template for the organized study of the phenomenon. Mack's research revealed an unusual mixture of physical and spiritual qualities woven through the abduction accounts. He recognized three categories of UFO/ET perceptions among his patients and other experiencers…

1. Evidence in the normal physical universe. [198]

- unusual lights
- radar blips

[198] You may notice these categories resemble Hynek's CE classifications in some ways.

- burned patches of earth

2. *Phenomena that appear to be explainable with physical laws if we had better technology.*
    - speeds and maneuvers that seem impossible with our technology
    - vanishing or appearing suddenly
    - telepathy with humans

3. *Observations for which we can conceive no explanation in the physical universe.*
    - "sense that experiences are not occurring in our space/time universe"
    - "a consciousness abductees experience of vast other realities"
    - "deeply felt sense of opening up to or returning to the source of being or creation"
    - "experience by abductees of a dual human/alien identity" [199]

Clearly, the phenomena experienced in this third category seem to have crossed into another realm. Once again, the experiences are qualitatively different from the former ones. Perhaps most puzzling is this abrupt change in mode. Certainly, the presumed ETs may do whatever they wish, but I believe we can honestly ask, why this switch? Apparently, there is no problem with the appearance of material crafts or their physical interactions. I could envision a scenario in which aliens continue the CE-3 contacts in nuts-and-bolts (or rather "skin-and-bones" or whatever) form.[200] In other words, there seems

[199] John E. Mack, *Abduction: Human Encounters with Aliens*, (New York: Charles Scribner's Sons, 1994), 31-32.

[200] I realize I'm speculating a lot here, knowing nothing about the alleged aliens. But I believe we can make some more or less educated guesses about them. My point is that

to be no logical reason to change "formats" in the middle of an abduction.

A Riddle, Wrapped in a Mystery, Inside a Spaceship

If the government and military seem "flighty," the ETs certainly aren't doing much better. Granted, there's a danger in attempting to anthropomorphize their actions, but I think we can apply some broad reasoning here. Are they looking for something? Are they just studying us? Vallee noted how absurd this curiosity/guinea pig theory is. The assumption is that to travel here in sophisticated spacecraft would take sophisticated technology. How long does it take mega-brains to study Earthlings? Shouldn't advanced space travelers have all the info they need by now? Even Hynek has stated…

> It seems ridiculous that any intelligence would come from such great [distances] to do reportedly stupid things like stopping cars and frightening people.[201]

Stories of forcible abductions and other physical manipulations seem to be almost as common as their fictional TV and movie depictions. Victims recall nightmarish memories of probings, surgeries, and other molestations. Yet despite what TV and tabloids promote, abductees rarely exhibit verifiable implants or surgical scars. Few if any plausible accounts involve what we might call real-world meetings or communication with a "tangible" being.

Along with many others, I've tried to imagine what visiting ETs are attempting to do or say. Like Hynek, I assume

---

they seem to prefer the physical universe for their travel (or at least that's the way they prefer to manifest their travel means). Our planes engage them. Radar detects them. Their movements sometimes are physically difficult to explain, yet still physical (point A to B, dodging, landing, etc.)

[201] Ian Ridpath, "Interview with J. Allen Hynek," *Nature*, October 4, 1975.

that if aliens are visiting us, they must be highly advanced to accomplish the feat. Furthermore, since we haven't all been zapped to ashes already, they must be either benevolent or indifferent. So, we would expect ETs to have already made some type of a worldwide, public debut.

But they haven't... When they make an appearance, it seems to be only to random individuals. Then maybe the messages ETs are leaving with these isolated abductees contain significant morsels of technology or wisdom to be passed on. That isn't the case either. Why can't they throw us a few bones?

Furthermore, aren't there better ways to send a message to humankind than by probing someone in a dream-state and leaving a few confusing memory riddles? If I were contacting another planet and didn't want a lot of publicity, I think I would just beam up a phone and call important people. Don't highly advanced creatures know how to hack into email or social media? Seriously, herein lies the conundrum. The enigmatic behavior of alleged ETs isn't consistent with what our commonsense demands. And I realize my reasoning here is more intuitive than scientific since I'm presupposing a lot about the nature of ETs. Nevertheless, considering their alleged visitations, wouldn't they probably operate on a few similar universals such as purpose, logic, industry, creativity, curiosity, and some semblance of sociability?

And back to the ETs' sage, civilization-altering wisdom... Aside from a few star charts and intergalactic anecdotes, ETs are quite monotonous. There's no useful medical, scientific, or technical advice. There are no practical ideas for improving society or a constitution for starting a "United Federation of Planets." A typical contact scenario resembles a medieval exorcism more than the meeting of two species. What happened to "Take me to your leader?" Their messages are more séance than science.

This has frustrated UFO aficionados for decades. By the 1970s some were already making the connections. John Keel noted…

> The endless messages from [ETs] would now fill a library, and while the communicators claim to represent some other world, the content of those messages are identical to the messages long received by mediums and mystics.[202]

It becomes obvious that the UFO/ET paradigm doesn't fit with modern science and logic. I'm not speaking of the technologies (unearthly physics, antigravity, etc.), but rather the overarching behavior or "psychology" of the phenomena. UFOs and ETs seem to be inconsistent, unpredictable, "flirty," inefficient in supposed objectives, and often contradictory. Some have offered explanations for this, which I believe at best are unsatisfying:

- ETs are stupid. (Not likely. How did they get here?)
- Perhaps there are several different races of ETs? (That would only multiply the chances of messing up. And surely one group would have an understandable plan.)
- ETs are so smart they just appear stupid. (Wish people would say that about me.)
- There is "method in their madness." The flirty, inconsistent behavior is a ploy for some hidden agenda. (If this is the case, their agenda must not be the expected "take over the earth," destroy the earth," "study the earth," or "save the earth.")

---

[202] John Keel, UFOs: Operation Trojan Horse, (New York: G.P. Putnam,1970).

We often hear the theme that ETs are our "space brothers" who only want to help us grow and evolve. If so, why are they so erratic in their methods? Perhaps they're observing us to learn better strategies for organization and effective communication. It has been suggested that our next message to space should be "Figure out what you're doing then come see us!"

The Dirty Little Secret of Abduction Cases

After the work of Mack, mental health professionals began dealing with experiencers and abductees on a more practical level. One of the most surprising findings was that the cases were clinically similar to issues brought about by occult encounters. As mentioned, some had already noticed this association as early as the 1960s…

> The UFO manifestations seem to be, by and large, merely minor variations of the age-old demonological phenomenon.[203]

Just as with demonic dreams and ghostly harassments, C-3 encounters and abductions seem to occur as much in the mind as in the material world. With the more extreme stories, in which a person claims to have been violently abducted, virtually all take place in eerie, other-worldly settings, producing suppressed, dreamlike memories in the victims. Computer scientist and astronomer Jacques Vallee did extensive UFO research with Allen Hynek in the mid-1900s. One of the icons of ufology (and an agnostic), Vallee became convinced of the "spiritual" nature of UFOs…

---

[203] John A. Keel, *UFOs: Operation Trojan Horse*, (New York: Putnam, 1970), 299.

> At first view, the very suggestion of the link is disturbing to a scientist… the phenomena reported by witnesses involve poltergeist effects, levitation, psychic control, healing and out-of-body experiences… familiar [in] occult literature.[204]

Bad news for naturalism… Indeed, naturalists would hope UFO phenomena are explicable in materialistic terms, even if this points to the possibility of sentient, extraterrestrial life. For the naturalist, a supernatural source would be a worst-case scenario, yet after decades of data, these implications remain.

Not only do ET encounters seem to mimic occult and spiritualist phenomena, the selection of participants is also heavily skewed toward those with previous ties to mysticism and the paranormal…

> In looking at the background of UFO abductees, it quickly becomes clear that almost to a man, they have some background in New Age or occultic beliefs. Interestingly, studies show that there are very few practicing Christians or Jews amongst UFO contactees. What could this mean? Are the aliens racists? Or does this, rather, indicate something about the belief systems of the abductees themselves?[205]

Certain researchers have concluded that the data show little difference among experiencers with respect to ethnicity, gender, age, and religion. Here is where surveys can be misleading. When data are analyzed more discriminately, an interesting fact is revealed. Referring to the abductee population, Bates explains…

---

[204] Jacques Vallee, *Messengers of Deception: UFO Contacts and Cults*, (Berkeley: Ronin, 1979), 204-5.
[205] Goldman, Stuart, *Christian Research Journal*, Summer 1990.

> …there is one group of people that, by and large, is notably absent. They are Christians… I am talking about what are known as "born-again," Bible-believing Christians—those who are often (these days unflatteringly) described as "Christian fundamentalists." It is as if ETs tend to avoid this select group of people.[206]

I certainly don't believe we should generalize from these statistics that *all* experiencers are unbelievers, nominal Christians, or occultists. Mature believers aren't immune to deception and spiritual pitfalls, but they do have armor and an Advocate. Bates is concerned that most churches aren't equipped to deal with abduction accounts because Christians generally avoid the topic. Immorality and substance abuse are approachable—claims of abduction are not.

One parachurch group has as its mission the investigation of ET phenomena and ministry to abductees and experiencers.[207] Their success rate in squelching the phenomena is significant. One of their more interesting findings is that when the name of Jesus is evoked during an alleged ET/abduction event, the results are the same as with demonic manifestations. The "ETs" seem to stop the experience and either vanish, fly away, or react wildly as fleeing demons. Obviously, accounts such as these don't get a lot of media attention.

---

[206] Gary Bates, *Alien Intrusion*, (Powder Springs, GA: Creation Book Publishers, 2018), 254.
[207] Gary Bates, "Lifting the Veil on the UFO Phenomenon," *Creation* 33(3):20–22, Jul 2011.

Are UFO and ET Phenomena Consistent with the Ephesians Hypothesis?

The title of this section asks a question that is either ridiculous in its credulity or profound in its insight. As with all sensory information that enters the human brain, UFO/ET/abduction data (first or secondhand) will be processed through an individual's worldview filter. I have biases, just like every other person who has ever lived. I make no excuses, nor do I offer any apologies. And to be clear, I don't expect my examples and reasoning here to convince the skeptic. My purpose isn't to lay out a scientific treatise, but rather to allow Christians to see that the same faith they have expressed in a supernatural salvation can also embrace the reality of the supernatural drama presented in Scripture.[208]

Yet I believe this is still a data-driven hypothesis in that it's consistent with the fundamental empirical evidence. This isn't simply playing the god-of-the-gaps card (or in this case demon-of-the-gaps). The Ephesians Hypothesis has explanatory and predictive power regarding the "big picture" of UFO phenomena. Hynek, Vallee, and many other secular researchers found no conflict between the data and this idea. There is no need to force-fit UFO and CE-3 accounts into a supernatural mold. (Be sure to note, I'm referring here to the 5%, or RUFOs, that seem to imply supernatural or extraterrestrial origin.[209]) For those willing to see, there is more than enough evidence to warrant the inclusion of this spiritual explanation for UFO/ET phenomena in the overall discussion. More recently, Christian scholar and pastor Mark Hitchcock makes the same connection as Vallee with a bit more specificity…

---

[208] This isn't to say that the Christian faith is without evidence. There are multitudes of Christian apologetics, both scientific and philosophical. Unlike many belief systems, the biblical worldview is supported by a *reasonable* faith.

[209] In no way do I contend that every odd, unexplained UFO incident has a demonic source. It would be unwarranted and irresponsible, as well as a bad testimony, to mark every anomaly as demonic.

> It is our belief that the UFO enigma is a demonically driven phenomenon. In our research of this subject, there is nothing in the realm of ufology that we don't find explained by the realm of the demonic. There is a consistent parallel between UFOs and the demonic/satanic that cannot be denied. There is a clear parallel between UFOs and the occult, as well as UFOs and the New Age movement. [210]

In another work, he notes a possible motivation for the phenomenon…

> Belief in UFOs fills many people's spiritual hunger with alleged higher sources of knowledge and experience. All of this is consistent with Satan's portrayal in Scripture as a masquerading deceiver…[211]

All things considered, I believe that a spiritual view of the 5% seems to provide coherent answers to the "why" question and other inconsistencies discussed earlier. Can we expand on Hitchcock's suggested motive? What would Satan have to gain by proliferating a mythology of extraterrestrials, and would this be within his ability? Let's examine some UFO/ET enigmas that are addressed by the Ephesians Hypothesis…

1. The RUFO phenomena, in general, are often elusive and inconsistent. We've alluded to this issue quite often, and it hovers in the background of almost every media account—UFO data give contradictory signals as to purpose. Do they wish to be seen or not? UFOs float slowly over heavily populated areas,

---

[210] Mark Hitchcock and Scot Overbey, *Extraterrestrials: What on Earth Is Going On?* (OKC: Hearthstone Publishing, Ltd., 1997), 112.
[211] Mark Hitchcock, *101 Answers to Questions About Satan, Demons, & Spiritual Warfare*, (Eugene, OR: Harvest House Publishers, 2014), 77.

low enough to pique curiosity, yet high enough to avoid detailed observation. Sometimes radar sees them—sometimes not a trace. They fly teasing close to our best military weapons, then vanish. When they make a close encounter, it's to a few who are unprepared or unbelieved. Are UFOs more theatric than technical? Vallee observed…

> Human beings are under the control of a strange force that bends them in absurd ways, forcing them to play a role in a bizarre game of deception.[212]

Moreover, are intelligent entities staging "just enough" at "just the right time?" UFO researcher Nick Redfern discusses this characteristic…

> There's the nature of the entities themselves: they practically overemphasize who, or what, they claim to be… We only have to take a careful look at such cases … to see that these "incidents" are clearly stage-managed. It is a game, a scenario that has *everything* to do with trying to emphasize the ET meme … Of course, they could easily avoid us! But, here's the deal: *they want to be seen*. It's not an accident. It's carefully planned. And it's designed to plant an image of "ET scientists" in the mind(s) of the witness(es).[213]

What if our confusion over UFO *behavior* stems from wrong-thinking about their *nature*? If the phenomenon has a spiritual source, the charades of Satan and his fallen angels don't necessarily need to be sensible, consistent, or believable. *The goal of the Father of Lies is to lead humans away from God's truth.*

---

[212] Jacques Vallee, *Messengers of Deception*, 20.
[213] Nick Redfern "UFOs: Extraterrestrial? Probably Not," www.mysterious-universe.org, 2016, Accessed 4 Jun 2020.

Deception, then, could use literally any strategy—silly, serious, simple, complex, enigmatic, believable, fearsome, scientific, fanciful, religious, secular, bizarre, classic, esoteric, logical, illogical, old-school, cutting-edge, mysterious, virtuous, or *anything that works*… Lewis warned that any demonic deceptions are possible, "provided that their cumulative effect is to edge the man away from the Light and out into the Nothing." [214]

2. Why so many different manifestations? It seems that UFO styles change with trends in earth technology. Ancient sightings were like angels or birds, with flapping wings and feathers. Accounts of strange aerial phenomena from the 1800s describe them as floating airships (blimps). A bit later they were gavotting as noisy "steampunkish" machines like those in Jules Verne stories. By the mid-20th century, observers saw mostly the iconic saucer form. Now, UFOs commonly appear as ethereal, non-physical craft. Also, in earlier close encounters ETs claimed they came from the moon, Mars, or Venus. Now that we know better, they hail from conveniently far-removed star systems.

Even though UFO/ET phenomena are erratic as to *motive*, they seem very predictable in terms of *mode*. It isn't the case 100% of the time, but their mode generally seems to coincide with the expectations of the observers…

> …the UFO phenomenon simply does not behave like extraterrestrial visitors. It actually molds itself in order to fit a given culture.[215]

---

[214] C.S. Lewis, *The Screwtape Letters,* (New York: HarperOne, reprint edition, 2015).
[215] John Ankerberg and John Weldon, *The Facts on UFOs and Other Supernatural Phenomena*, (Eugene, OR: Harvest House, 1992), 10.

Interestingly, Keel made the same observation about paranormal phenomena in general…

> If we are young and living in a Catholic community, It becomes a BVM [blessed Virgin Mary]. If we are atheistic and immersed in science and technology, It appears in a mechanical-looking contrivance in the guise of a being from some distant galaxy.[216]

Some attribute this to the gullibility of observers, but we've already dealt with that aspect in designating the RUFOs. This becomes almost a clarion cliché by its ubiquity: *We can't dismiss as gullibility, hoax, or recklessness such a significant number of accounts given by reliable witnesses*. So, does a spiritual cause have explanatory power in this aspect, as well? Wouldn't a deceptive Enemy use the preexisting beliefs of his targets?

3. Do they want to kill us, save us, use us, or probe us? Alleged messages and abduction experiences are diverse and contradictory, but the two central teachings are 1) a vague New Age pantheism, or 2) the atheistic idea that they (the ETs) are somehow responsible for the existence of humankind on Earth. I'm not arguing here whether these ideas are true or false. Rather, I'm perplexed by the odd mismatch of behavior and message if ETs are indeed behind the phenomenon. Ankerberg states this well…

> …how credible is it to think that literally thousands of genuine extraterrestrials would fly millions or billions of light years simply to teach New Age philosophy, deny Christianity, and support the occult? Why would they do this with the

---

[216] John Keel, *The Complete Guide to Mysterious Beings,* (CreateSpace Platform, rev. 2016), 217.

preponderance of such activity already occurring on this planet? [217]

Hitchcock supports this observation…

> The curious thing is that the theology [of ETs] is New Age theology; namely, that we are all divine, all evolving to higher levels of consciousness, that reality is what you perceive it to be, etc. These are anti-Christian teachings. In some cases, comments about Jesus himself have been made that directly contradict biblical revelation. I've studied the New Age for quite some time, and a great many people in the New Age not only accept aliens and UFOs but teach that they are more highly evolved than we are; and that we need to learn from them.[218]

Another prevalent theme of ETs' alleged communication is that their purpose is to "help" humans. However, this help is usually nebulous, generalized advice, much like a daily horoscope. One would think that beings with the intelligence and technology to traverse light-years could give us a bit more. Perhaps I'm being too pragmatic, but it would be nice to get a cure for cancer or a cheap, clean energy source. And if their assistance must be limited to the moral and spiritual, why not give some specifics about how to reach the "higher plane" they preach? At least Yoda taught Luke how to levitate a spacecraft and wield a lightsaber.

I make no claim that any of this is strong evidence against the ETH. I only point out that these messages are highly consistent with messages we might expect from demonic

---

[217] John Ankerberg and John Weldon, *The Facts on UFOs and Other Supernatural Phenomena*, (Eugene, OR: Harvest House, 1992), 13.
[218] Mark Hitchcock and Scot Overbey, *Extraterrestrials: What on Earth Is Going On?* (OKC: Hearthstone Publishing, Ltd., 1997).

sources. They're virtually indistinguishable from the signature information we get from psychics and occult practitioners.

In addition to giving us the battle strategy for fallen angels, Ephesians also identifies another of Satan's titles. Paul reminds the Ephesian Church of their former trajectory of "following the *prince of the power of the air*, the spirit that is now at work in the sons of disobedience."[219]

Some connect the Deceiver's role, "Prince of the Power of the Air," directly to UFO phenomena. It's a bit of a stretch exegetically, but if Satan does presently claim this domain of Earth as his own, it's no wonder he uses it for his dark purposes.

Is this all too broad a mental jump? Our 21st-century minds often have trouble wrapping a modern phenomenon in what culture wants to keep a medieval myth. I believe even this could be part of the deceptive strategy. In upcoming chapters, we'll examine some details of how this deception might play out.

---

[219] Ephesians 2:2

16
A Paranormal Potpourri

"I've seen movements in the clearing.
Someone sent you something satanic..."
Sir Elton John[220]

"Big deal! The devil can move something.
The devil can tell you something.
The devil can do something unusual.
Why should we be shook up about things like this?"
Walter Martin[221]

Once while standing in line at the grocery checkout, my attention was drawn to the elderly couple in front of me. I stepped closer to make sure… Yes, I was correct. They both had prosthetic noses. I had never seen a prosthetic nose before, and certainly seeing one would not have been highly unusual. But two… On two apparently married seniors?

In no way am I minimizing this. The loss of a nose is undoubtedly a tragic and miserable occurrence. But I couldn't help wondering… "What causal factor would have resulted in two closely-associated people experiencing such a rare event?"

---

220 Elton John, "I've Seen the Saucers."
221 Walter Martin, "Tools of the Occult," Audio on www.calvarytruths.com, Accessed 23 May 2020.

As I inched forward in line, several hypotheses came to mind. Did they both share a degenerative sinus infection? Did they both inspire radioactive particles? Was there an inhumane attack of some sort? Were both thrown through the same windshield? Did they meet at a support group for those who had lost noses? To this day, my hypotheses remain untested, even though I've wasted way too much time thinking about them.

We've seen that paranormal phenomena are often concentrated around certain places or *hotspots*. Two or more similar incidents in a non-random relationship imply another level of significance. These clustered data can be misleading in that it's sometimes difficult to distinguish cause and effect. Are multiple phenomena *drawn* to the location, or does something about the location *cause* the phenomena?

For example, ufologists have noted that UFO sightings seem disproportionately higher near military bases. Several logical *causal agents* could be put forth. Despite the official spin, perhaps some UFOs really are top-secret military craft. Or do ETs simply have an interest in the military? For now, let's not be concerned with the correctness of either explanation. Just note that our usual assumption is that data clusters have a relational (non-random) cause.

There are other *nuisance variables* that might come into play. Could the people at a particular location be more likely to see or report phenomena? Does the geography of a location simply make observation easier? Is it possible that reports of an initial event could trigger similar ones? Or perhaps more controversial, does the local culture of an area make expecting, imagining, observing, or reporting phenomena more likely? Here again, we must often apply the Pickle Principle mercilessly to weed out the non-genuine.

Some paranormal hotspots possess an additional complexity. Let's call the previous type of data cluster a *single-phenomenon* (SP) hotspot. What are we to make, then, of *multi-phenomenon* (MP) hotspots? How does the logic change when at one location, several seemingly *different types* of paranormal phenomena are claimed?

Vallee was one of the first to draw attention to MP hotspots…

> It is the rule, rather than the exception, to find significant UFO sightings preceded or followed by other anomalies, notably of the poltergeist variety.[222]

UFOs and ghosts? Are we stuck in a late-night B-movie? The uncomfortable truth is that Vallee was correct and similar combinations still haunt paranormal investigators.

We might expect combinations of these incidents and variables to be more difficult to analyze than the individual phenomena. However, I believe their symbiosis provides a valuable clue in analyzing paranormal phenomena. First, let's look at some specific examples.

Bridgewater Triangle

Had the Pilgrims only known about the area just west of Plymouth, would history have been different? The Bridgewater Triangle has generated enough paranormal lore to knock the knees of the staunchest settler. Bounded by the Massachusetts towns of Abington, Freetown, and Rehoboth, the 200-square-mile triangle has a long history as a multi-phenomenon hotspot. Some ghostbusters trace the strangeness to Native American curses and spirits. In the 1670s, relationships between the colonists and the Wampanoag tribe deteriorated, leading to numerous skirmishes in what has been named King Philip's War. Many historians cite this as one of the bloodiest wars on

[222] Jacques Vallee, *Confrontations: A Scientist's Search for Alien Contact,* (New York: Ballantine Books, 1991).

American soil. Wampanoag legend says that the land would be under a curse until the lost belt of King Philip (their chief) is returned to the tribe.

The Wampanoag occupied the 16,000-acre Hockomock Swamp as a launching point for attacks on the colonists. Several tribes have used it as a sacred burial site. Aptly, "Hockamock" is Algonquin for "the place where spirits live." Roughly at the center of the triangle, the Hockomock Swamp seems to be ground-zero for the region's paranormal activity. The swamp is allegedly home for a variety of cryptids, including Sasquatch, giant wolves, wolfmen, and giant cats. There have also been documented (by police) reports of large flying creatures resembling pterodactyls, or *thunderbirds*. Adding to the mystery, many have seen huge snakes "as thick as a stovepipe" in the swamp.

Ghost and poltergeist claims are commonplace in the Bridgewater Triangle. The most popular recent apparition is the "red-haired hitchhiker" seen by drivers on Route 44. The middle-aged man is easy to identify by his black eyes and tendency to vanish when offered a ride. Current residents still see the elusive *pukwudgies* who have vexed Wampanoags for centuries. These four-feet-tall, humanoid creatures have smooth, gray skin and almond eyes (sounds like a Gray, huh?) and have the ability to fade away suddenly, hence the name translation "little wild man of the woods that vanishes." It's interesting that encounters with the creatures are almost always accompanied by glowing orbs (*tei pai wankas*) and a large, shadowy form not far away. When the gospel reached the Wampanoags, the pukwudgies were identified with demons, still set on frightening, harming, and destroying.

With the parade of other oddities, it's easy to overlook the UFO sightings over the Bridgewater Triangle. Citizens claim that UFOs actually dominate the paranormal reports in the area. Orbs and other lights seem to be the most common. On Halloween night 1908, before the modern age of flying saucers, many in Bridgewater witnessed a large glowing "lantern" navigating over the town. The most extensive UFO flap in the history of Massachusetts occurred in 1979. Several videos exist showing nighttime orbs, exhibiting unusual flight behavior. The best-documented incident was a low-flying, shield-shaped craft over Raynam, estimated to be larger than a football field.

Any discussion of the Bridgewater Triangle phenomena must include the occultic activity in the area. At the southern end of the triangle, in and around the Freetown State Forest is a different type of hotspot—an epicenter of crime and evil. Beginning in the 1980s, the triangle experienced a high proportion of crime and violence including occult activity, organized crime, murders, suicide, drugs, and general violence. Despite dissociation by the Church of Satan, several sites for satanic rituals were verified by police. Animal mutilations, sacrifices, and at least one ritualistic murder were attributed to satanic cults operating in there.

Some have attempted to link these evils to the paranormal mystique of the location. It would be difficult to prove such a connection. Cause and effect factors can be misleading. Considering only the occult influence in the area, do places with a rich paranormal tradition tend to attract those with occult views? If some paranormal phenomena have demonic sources, could the area be predisposed to other types of evil? Is this evidence of territorial spirits (demons)?

Most regions have their own folklore. Paranormal researchers are quick to note that the Bridgewater Triangle phenomena are qualitatively different in preponderance, diversity, and number of reliable witnesses (not to say that many aren't hoaxes and silliness). There is little indication that this hotspot will fade from memory anytime soon.

Skinwalker Ranch

Skinwalker Ranch has become perhaps the most studied multi-phenomenon hotspot on the globe. The 500-acre plot in northeastern Utah is allegedly a showcase of paranormal events. UFO reports have been common in this Uintah Basin region since the 1970s. Accounts of other paranormal phenomena can be traced back hundreds of years in folklore and Native American legends.

In 1995, the Sherman family bought the acreage and began cattle ranching. After being on the property for under two years, the family called it quits due to what some referred to as a "paranormal Disneyland." George Knapp who has written extensively about the ranch describes it…

> …as if some cosmic puppet master had written a laundry list of every spooky phenomenon of modern times and then unleashed them all in a single location, resulting in a supernatural smorgasbord that no one could possibly believe, even less understand.[223]

The list of odd occurrences included…

- UFOs – Both lights and nuts-and-bolts, many small glowing orbs.
- Poltergeist activity – Numerous items disappearing and reappearing hours or days later, stacks of wood moved across a field, farm equipment placed high in treetops.
- Cattle mutilations – Organs removed with surgical precision, no blood at site.
- Cryptids – Super-sized wolves, wolf-like humanoids, and other creatures.

---

[223] Colm A. Helleher and George Knapp, *Hunt for the Skinwalker: Science Confronts the Unexplained at a Remote Ranch in Utah*, (New York: Paraview Pocket Books, ebook, 2005).

- Phantom holes – Large amounts of soil disappearing leaving craters.
- Odd energy effects – Magnetized corrals, frozen engines, radiation burns, unusual behavior of electronic devices.
- Men in Black – What paranormal adventure would be complete without these mysterious observers?

The Shermans' fear of being stuck on cursed land was assuaged in 1996 when Robert Bigelow purchased the ranch. Among many other entrepreneurial successes, he is the founder of Bigelow Aeronautics and well-known for his interests in UFOs and other paranormal phenomena.[224]

Bigelow immediately made large investments in the property, setting up elaborate surveillance systems and bringing in another of his pursuits, the National Institute for Discovery Science (NIDS). He established the organization as a kind of SWAT team to bring together credible scientists for paranormal research. NIDS scientists and security officers monitored Skinwalker Ranch 24/7 for months, reporting some of the same phenomena as had the Shermans.

The results of the observations were frustrating according to Bigelow and the NIDS leaders. Despite the high-tech approach, tangible evidence was elusive. The scientists and crew claim to have witnessed numerous paranormal incidents but were unable to gather the solid scientific proof they sought. The consensus of the NIDS scientists was that the phenomena were "random and unpredictable, and never happened more

---

[224] Robert Bigelow has numerous paranormal connections. He not only established NIDS, Bigelow Aerospace is involved in several far-reaching space projects. Bigelow was a major force in lobbying for the AATIP program. He also states that he is "absolutely convinced" that ETs have visited earth. (Helene Cooper, et al, "Glowing Auras and 'Black Money': The Pentagon's Mysterious U.F.O. Program," www.nytimes.com, 16 Dec 2017.)

than once in the same place or in the same way."[225] Paraphrasing the researchers: "It's like they know what we are planning and always try to mess up our investigations, and prevent the collection of solid evidence."

In 2016, Utah real estate tycoon Brandon Fugal purchased Skinwalker Ranch. Paranormal investigations are on-going with a new team of scientists.

Dancing with the Apparitions

The Bridgewater Triangle and Skinwalker Ranch phenomena present some unique interpretive challenges. Analyzing a single-phenomenon site is often difficult. Analyzing a multi-phenomenon hotspot is orders of magnitude more complex.

If I have five symptoms of illness, I don't assume I have five different causal agents (although that *could* be possible). Unless sicknesses are the daily norm for me, my assumption should be that simultaneous multiple symptoms stem from a common cause. Knapp notes this principle…

> Which particular research data from the Utah ranch does one have to discard in order to fit the rest into a tidy box? Or should we just ignore all of the tidy boxes?[226]

Let's go ahead and state the obvious issue with Skinwalker Ranch and other MP hotspots: *Observing several different paranormal phenomena in close association is crazy* (or some*one* is crazy)…

If I went to the ER with a gunshot wound to my pinky toe, claiming a freak accident, medical staff would fix my toe

---

[225] Helleher and Knapp, *Hunt.*
[226] Helleher and Knapp, *Hunt.*

(maybe with raised eyebrows) and send me home. If I came in with *both* pinky toes shot, claiming two different accidents, my toes would be fixed, then I would be transferred to—well, a different wing of the hospital. Logic demands a commonality between associated, random, low-probability events, and in this case, my mental state would be the suspected connection.

Are we misinterpreting one or more of the "tidy boxes" encompassing MP hotspots? Despite what you may have experienced in high school chemistry, empirical science loves simplicity. Back to the disease example, it's much more *parsimonious* to assume one common cause for five simultaneous symptoms. In other words, the *principle of parsimony* (Ockham's Razor), says that the simplest solution is probably the correct one.[227] In the quest for more and more data, good scientists always strive to organize the data and discover broad over-arching explanations that can provide meaning to apparent randomness.[228] Currently, for example, physicists are seeking a *theory of everything* (ToE) that would tie several seemingly unrelated physical laws into a simpler model. None of the quantum, string, M, gravity, or relativity boxes seem tidy enough.

So, what is the most parsimonious explanation for paranormal hotspots? Is there a commonality that could ease the "craziness" problem? What are some possible connections among ufology, ghost-busting, and cryptozoology at Skinwalker Ranch? In a discussion of MP hotspots, Keel concludes…

---

[227] Obviously, not always the case. There are times when the answer is more complicated than first thought. But it's acceptable to use Ockham's Razor as a logical starting point.

[228] Scientists use *inductive reasoning* in observing data and creating a hypothesis. (Owls fly, robins fly, sparrows fly… My hypothesis: Birds fly.) Then using *deductive reasoning*, a hypothesis is tested. (Birds fly… If that's true, then that unknown bird sitting on the branch should be able to fly.)

> … many of these events may have been somehow interrelated and may, when viewed in toto, represent a larger whole. Witnesses of hairy monsters and witnesses to UFO events and psychic manifestations have experienced many of the same things. Instead of dealing with mere extraterrestrial visitations… we may be dealing with brief glimpses of something far broader and infinitely more complex.[229]

There is a tendency to place the UFO component at the epicenter of multi-phenomenon activity, likely because UFOs are arguably "less crazy" than other paranormal alternatives. I suppose I would rather be psyched by an alien than by a ghost. But as we might expect, worldview is a strong factor in weighing hypotheses and identifying causal agents…

1. A ghost hunter might explain the Skinwalker Ranch phenomena as "Spirits of early people killed on the land are haunting it. They are appearing as UFOs, poltergeists, and strange beasts in order to drive away human intruders."[230]

2. Ufologists could claim "Aliens are using the area for base operations. They employ advanced theatrics to confuse and scare away humans."

3. Cryptozoologists have an answer, too. "A population of highly-advanced, secretive werewolves has lived in the area for millennia. They wish to remain unknown and are attempting to discourage human encroachment."

---

[229] John Keel, *The Complete Guide to Mysterious Beings,* (CreateSpace Platform, rev. 2016), 156.
[230] These different viewpoints are examples of how different *worldviews* might interpret the same event (see Appendix B).

4. *New Agers* can explain everything easily. "Gaia, the Earth Spirit, is unbalanced because man desecrated the land. She is fighting back with all manner of psychic manifestations."

5. *Skeptical naturalists* might accuse: "It's one big hoax. Everyone is in on it—NIDS, the ranchers, the Utes, the neighbors, and the government. It's a publicity stunt for books and movies."

6. *Parascientists* theorize: "Perhaps quantum tunneling is allowing a portal to form between two dimensions. A variety of vehicles and beings are accidentally coming through from all over the galaxy, and they are as confused as we are."

7. *And, of course, the government* can't admit it, but they too have a hidden agenda: "One of our top-level branches is involved. We can neither confirm nor deny what they are doing, but suffice it to say, it's important for national security."

I would guess there are psychologists (both professional and armchair) who would like to weigh in, as well. It would certainly simplify the whole business of dealing with the paranormal if we could give a blanket-diagnosis of insanity (or maybe just a touch of neurosis) for all experiencers. Undoubtedly, mental aberrations (whether pathological or just over-active imaginations) account for some paranormal claims. No one disputes this. But the preponderance of reports from many reputable observers has always been the fly-in-the-ointment for skeptics. Remember, a premise of this book is that in a proportion of paranormal incidents, observers may be making some *observationally accurate* observations.

The only thing worse than having the wrong view of something is to have a mixture of wrong views. We've discussed how some Christians compartmentalize their views regarding spiritual matters—"What happens on Sunday stays on Sunday." Postmodern thought is good at this *pragmatic hypocrisy*. There is a disconcerting trend today to create a separate compartment (or *tidy box*) for each supernatural phenomenon.

Just this week, I listened to a podcast with a published paranormal researcher who exemplifies this. From his Christian background, he believes that angels and demons exist and that demons may possess some people. But he also believes that ETs are operating on Earth and that it may be difficult at times to distinguish whether a person who claims to be an abductee is relating a demonic experience or an alien encounter. This investigator also believes that it may be possible for a ghost (the spirit of a dead human) to mimic both demonic and ET phenomena. But wait… He also speculates that evil beings may be entering our realm through interdimensional portals. At a recent seminar, he outlined ways of determining which of these four different "paranormal" entities might be harassing a troubled individual.

In a way, this is a form of paranormal "polytheism," (or at best dualism) and it's disturbing that many Christians buy into it. It's as if God says, "I'll deal with the demons, but those other phenomena—they're out of my jurisdiction!" Not only does this concept snub Ockham's Razor, but it also mocks the revealed nature of the Most High. Kuyper reminds us of the biblical truth…

> There is not a square inch in the whole domain of our human existence over which Christ, who is Sovereign over all, does not cry, "Mine!" [231]

Pardon the preachiness, but this touches on the main point of this book. Yahweh God is sovereign over all existing beings,

---

[231] Abraham Kuyper, From *Abraham Kuyper: A Centennial Reader*, ed. James D. Bratt (Eerdmans, 1998), 488.

locations, realities, and belief systems.[232] Restating an earlier idea, *if a supernatural source is indicated in a particular phenomenon, it will necessarily and singularly be of the reality described in Ephesians 6.* There are no supernatural alternatives to the one supernatural reality presented in Scripture.

It bears repeating at least one more time: *Worldviews have consequences...* And I believe the biblical worldview best provides a coherent interpretation of activities beyond our senses. This isn't a new or radical concept, but rather the traditional, scriptural view of reality. The Ephesians Hypothesis has explanatory power not just in paranormal phenomena in general, but in multi-phenomenon cases. None of the other proposed explanations withstand scientific, philosophical, or evidentiary scrutiny. That was the conclusion of NIDS and others who have investigated the phenomena.

If these incidents are genuine, the random combination of paranormal components appears to be too bizarre to apply the usual SP explanations. They point to a more purposeful and orchestrated phenomenon. (And that is my assumption—logical connections among all paranormal phenomena.) Unlike other alternatives, the Ephesians Hypothesis is as robust in explaining MP activities as it is with SP ones. In fact, I believe the Enemy may be inadvertently "showing his hand" in MP manifestations. *When craziness is pushed too far, it draws attention.*

Christians should not be shamed into silence simply because the opinion police attempt to squelch anything with biblical connections. Some from an opposing worldview might still argue that this hypothesis works only because it presupposes evil supernatural entities with great power and a

---

[232] Indeed, good news for the believer who is truly in Christ. There is only one supernatural threat and Jesus, by his finished work on the cross, has "disarmed the rulers and authorities and put them to open shame, by triumphing over them..."

bent toward trickery and deception. To this I would say "Good! You're beginning to comprehend the biblical worldview!"

This is unquestionably a faith issue to a large extent, and if one wishes to leave it at that, the Ephesians Hypothesis is as good an explanation as the Scripture is trustworthy. But for those who need a more tangible base, we can dig deeper. (I'm so glad Jesus didn't condemn Thomas for this stance. When Thomas the Curious needed empirical data, Jesus readily offered him palpable, physical evidence.)

The Quintessential Ghost

This is possibly one of the most important sections of the book. Thus far, we've looked at various phenomena individually and as they occur in some MP cases. We've also applied the Ephesians Hypothesis in a piecemeal fashion as a potential explanation for some paranormal events. I think it's time to pull it all together into a *tidier box* or even something close to a supernatural *theory of everything*.

I admit there is danger here. Even Ockham would confess that his razor cuts both ways. By making things too parsimonious, we run the risk of missing some causal factors. In other words, we must not oversimplify at the price of sacrificing accuracy.

We'll begin with an individual phenomenon that I believe fits the Ephesians Hypothesis most comfortably—ghosts. Let's examine the details of the general theory of "ghosts" we arrived at earlier. (I assume most Bible-believing Christians accept that any genuinely paranormal "ghost" is actually a fallen spirit engaging in some form of deception, but for now, let's play along with the secular ghost idea.)

Tyrrell, who gave us the criteria for a "perfect ghost," described the paradox of apparitions he noticed in his research…

> Apparitions, then, combine two qualities. They are: a) non-physical in character, yet, when at their best, they are b) indistinguishable from material figures normally perceived, so far as the visual and auditory senses are concerned… it is not uncommon for the sense of touch to be hallucinated in apparitional cases.[233]

He also noted some specific abilities of a typical apparition as witnesses report them...

- The apparition can externalize in different ways. It might be perceived as ethereal or as solid as real matter.
- It has no actual physical basis, since it may vanish or pass through walls even after being sensed as solid.
- The apparition can imitate normal perception. It can fit into the surroundings and respond to them.
- It can externalize additional features other than its central figure, such as clothing, jewelry, furniture, carriages.
- The apparition can make itself perceived in the same way (or different ways) by multiple observers.
- It can manipulate perceptions of general surroundings such as air temperature, wind, clouds, precipitation, and lighting.
- It can alter the observers' subjective feelings to some extent, causing unwarranted fear, peace, disorientation, happiness, or sadness.

Now let's rough out a "ghostly phenomenon hypothesis" that might fit the Ephesians Hypothesis:

---

[233] G.N.M. Tyrrell, *Apparitions,* (New York: Collier, 1953, e-book 2018).

Satan's scripturally revealed aim is to deceive, confuse, and oppose all that is of God. It is within his ability, and that of demon spirits, to manipulate diverse phenomena either physically or through human perceptions. They have as their goal the age-old, continuing strategy of encouraging belief in a false supernatural and drawing humans away from biblical truth by whatever methods work.

How closely do the observations of ghost phenomena fit the Ephesians Hypothesis? Let's take a detailed look at ghost incidents as a model for other paranormal phenomena.[234] I've aligned the observations with the appropriate components…

1. There is only one supernatural reality—the unseen spirit realm described in Scripture. If the supernatural is indicated, it will be of this type, and this type only.
Observation: Ghosts seem to be spirits, but they don't necessarily have to be the spirits of dead humans or any other type of paranormal being.

2. Paranormal events will never be neutral. They are designed to lead the observer.
Observation: Ghost phenomena change people and culture. In most cases, the change is negative, causing fear instead of faith, or at best, a distorted worldview.

3. Biblical spirit beings can interact with the physical world.
Observation: Ghost phenomena seem to exist in the physical realm.

4. Fallen spirits are able to "lead" the thoughts of humans, as allowed by God.

---

[234] I encourage the reader to follow this model in analyzing any paranormal phenomenon.

Observation: Reports of ghost encounters indicate that witnesses seem to experience non-physical objects and entities. It isn't clear whether the phenomena are composed of matter and energy at the time of occurrence or are deceptive perceptions created in the observer's mind.

5. Not all evil is directly caused by fallen spirits. Its source can be our own fleshly nature and the evil of the world system.
Observation: Despite the number of unexplainable phenomena, many apparitions have natural explanations (hoaxes, false perceptions, overactive imagination).

6. Satan, the leader of fallen spirits, has the goal of opposing God in any way possible.
Observation: Except for possibly affirming the afterlife, apparitions contradict the biblical worldview in message and behavior.

7. Satan's chief strategies are lies and deception.
Observation: Superficially, ghost phenomena are generally confusing and enigmatic. Among incidents, ghosts provide contradictory information about the afterlife, God, and the nature of reality.

8. Satan and fallen spirits will use any means possible to deceive humans and lead them away from God and truth. In most cases, it's in Satan's interest to not reveal his true identity. Demons will manifest themselves to some, but maybe not others.
Observation: Ghost phenomena are quite effective in leading observers to false truth. The elusive nature of apparitions seems to be clear evidence of these strategies.

9. Satan can appear as an "angel of light."
Observation: Many apparitions claim to be benevolent. Even though their doctrines are false, the messages support a vague view of pseudo-Christian concepts, such as heaven, an afterlife, and a god-like being. Some ghost manifestations are literally "angelic" in nature.

10. Fallen spirits have some ability to manipulate matter and energy. Although Satan and fallen spirits have great power, the ultimate power is God.
Observation: Apparitions often include poltergeist phenomena. The actions are sometimes difficult to explain except as a spiritual interaction. They are limited and fleeting.

We don't always understand God's operations in the physical realm. Philosophers ponder why bad things happen to good people, and vice versa. God in his love and omniscience often allows a temporary *lesser evil* in order to gain an eternal *greater good*—for example, in his revealing of sin, leading to repentance, or sanctifying believers. It's difficult to comprehend, much less accept, that God allows Satan and his followers to exert any influence on this world at all, yet that is a biblical fact. Why would God allow paranormal events when he clearly has the ability to prevent them? We must assume the same answer as with any trial or tragedy: "For those who love God all things work together for good—for those who are called according to his purpose."[235]

If a suspected paranormal incident has a demonic source (or any other), it didn't escape God's notice. Whether it's the result of the general fallenness of the world, malicious individuals, or direct demon activity, God remains in control of the outcome. I believe, however, that we are accountable for our response to any danger or evil. Scripture lays out the divine strategy for confronting the supernatural.

---

[235] Romans 8:28

Ephesians urges us to put on the gospel armor so that we can "stand against the *schemes* of the devil." The word "schemes" is the Greek *methodeía* which has in mind an organized, methodical evil act that is well-planned. Some translations use "wiles" as in Wile E. Coyote. As everyone knows, Coyote never simply waits around for Roadrunner to happen by. He never recklessly chases Roadrunner. No, Wile E. has an arsenal of tools and highly orchestrated *methodeia.*

Perhaps a more sophisticated picture of *methodeia* is in the old *Mission: Impossible* series. The best part of each episode was watching the team put together the minute details of the deception, not knowing how all the random pieces fit. Then, in the last three minutes, it all comes together, and the trap closes on the evildoers.

It would be *extremely* naïve to think that Satan and his cohorts are haphazard or flippant in their scheming. When the Evil One engages in battle, *it's always the best possible strategy he can muster, and it's custom-designed to deceive his victims.* Satan always plays to win…

> The devil's brain is prolific with plans. He has many ways of doing many things… He never runs in ruts, Fruitful, diverse, and ever fresh is his way of doing things… He acts by trickery, and always by guile.[236]

Now let's take a big leap and apply this same reasoning to other paranormal phenomena. Is it apparent by now, that *if* ghosts did exist in the way secular culture believes, *any other paranormal phenomenon could be perpetrated by an alleged "ghost."*

If a ghost can cause illusions of a deceased person, couldn't it cause the illusion of a flying saucer or ET? If a ghost

---

[236] E.M. Bounds, *Guide to Spiritual Warfare*, (New Kensington, PA: Whitaker House, 1984), 106.

can provide obscure unknown knowledge, couldn't it pose as a psychic entity or ascended master? Please note: I'm not suggesting ghosts are responsible for all paranormal phenomena. Rather, I'm emphasizing that ghosts and all types of *genuinely* paranormal phenomena are of the same source—fallen entities from the supernatural realm. *I believe the biblically-defined spiritual reality is sufficient to explain all truly paranormal phenomena.*

Given this perspective, we may now ask whether individual paranormal events can be similarly explained. Even from a secular view, Hynek connected *close encounters* with *ghost encounters.* He concluded that UFOs and ETs might operate...

> ... in the same way that a poltergeist can produce very real physical effects.[237]

So, are UFO/ET phenomena material or mental? Walter Martin suggests...

> It seems evident that these phenomena are produced in the same manner that other occult phenomena are produced. They involve dramatic manipulations of matter and energy. Although they originate from the spiritual world, they can produce very powerful, temporary physical manifestations at the material level.[238]

We could conclude that paranormal phenomena could be either material or mental, in the same way that a ghost may be either.[239] The same hypothesis *may* be applied to manifestations

---

[237] J. Allen Hynek and Jacques Vallee, *The Edge of Reality: A Progress Report on Unidentified Flying Objects*, (Chicago: Henry Regnery, 1975), 262.
[238] John Ankerberg and John Weldon, *The Facts on UFOs and Other Supernatural Phenomena*, (Eugene, OR: Harvest House, 1992, ebook).
[239] And ultimately, it doesn't matter to the Enemy as long as his victims perceive the realness.

of spirit guides, psychic messages, folklore creatures, or any other paranormal entities. Whatever their form, their singular goal is to demean the place of humans in God's plan, produce fear, and replaces dependence on God with dependence on a false source of truth and power. Thus, our formal hypothesis of paranormal phenomena:

> *Satan's scripturally revealed aim is to deceive, confuse, and oppose all that is of God. It is within the ability of demon spirits to manipulate the diverse phenomena observed at some MP hotspots, either physically or through perceptions. The commonality is simple—fallen, rebellious beings with the age-old, continuing strategy of encouraging belief in a false supernatural and drawing humans away from biblical truth by whatever methods work.*

17
Could It Be Magic?

Bad reasoning as well as good reasoning is possible; and this fact is the foundation of the practical side of logic.
Charles Sanders Peirce[240]

"Sleight of hand and twist of fate..."
U2[241]

In Mark Twain's *A Connecticut Yankee in King Arthur's Court*, time traveler Hank Morgan wakes to find himself back in the Middle Ages. The classic plot has Morgan using his knowledge of science to amaze the citizenry with "magical" powers. To the dismay of the wizard Merlin, Morgan's science-based demonstrations convince the people that he is the superior wizard.

This type of *natural magic* attributes its power to some universal, impersonal force. Clearly, the force in this story was simply the laws of science applied at the right place and time. Since these were unknown to the primitive people, it would naturally be considered paranormal to them.

The magic depicted in J.K. Rowling's *Harry Potter* stories seems to be of the same type. Wizards (as opposed to non-

---

[240] Charles Sanders Peirce, "The Fixation of Belief," *Popular Science Monthly*, 12, November 1877.
[241] "With or Without You," U2, 1987, Island Records.

magical Muggles) are born with the innate ability to alter the physical world around them through wands, incantations, and spells. Their magic operates through very specific principles much like science and can be manipulated by the user within these bounds.

In another type of sorcery called *personal magic,* the practitioner believes they are summoning their power from an intelligent, supernatural entity. Purveyors of personal magic believe that by following the required procedures they are summoning the assistance of gods, angels, demons, dead spirits, ancestors, spirit guides, ascended masters, or even aliens.

In both methods, the magician is attempting to use supernatural sources to affect a situation in the physical world. This sounds very much like the description of a *miracle*. We might now further refine our definitions within a Christian context: A *miracle* involves God as the supernatural source; *magic* has a supernatural source other than God.

Magic in the Bible

Sorcery was still widespread during the time of the early church. Acts 8 tells of a magician named Simon who was operating in Samaria. He was one of several Jewish sorcerers who capitalized on his ethnicity and knowledge of Jewish traditions to mesmerize pagans. Under the teaching of the Apostles, Simon became a believer but had trouble forsaking his magic persona. He was enthralled by the Apostles' miracles and offered to pay for a quick training session. Long story short—Peter made it clear that there was no room for Simon's magic in the life of a Christian.

Paul and Barnabas encountered another magician on the island of Cyprus. Elymas the sorcerer "opposed them and tried to turn the proconsul from the faith."[242] Paul was quick to discern that Elymas was "a child of the devil and an enemy of everything that is right." Interestingly, the term *sorcerer* here in Greek is *magos*, the same term used for the magi who came to the infant Jesus.

Serious Magic

Many terms are under the umbrella of magic including sorcery, augury, charms, enchantments, curses, omens, divination, and necromancy. We often smirk at these old-fashioned, fairy tale notions of magic, but the Bible takes a much more serious stance. God made it very clear that his people must avoid even the most trivial suggestion of magic, and for good reason.[243] Whether real or imagined, any dependence on magic is a denial of God's ability or wisdom. Surrounded by pagan peoples who practiced magic as a lifestyle, the Israelites were always vulnerable.

I believe this was more than simply warning them against wasting their time on superstitions and folklore. Perhaps of all paranormal activities, magic is the most blatantly "dark-side" supernatural. Can you imagine walking into an auto dealership and asking, "Does anyone here think I need a new car?" It must be a great day in the underworld when one of their "clients" *asks* for magical assistance. Far from being humorous, dabbling in magic is a serious matter to God. Perhaps that's why he gives one of his strictest warnings to his children…

> When you come into the land that the LORD your God is giving you, you shall not learn to follow the abominable practices of those nations. There shall not be found among you anyone who burns his son or his daughter as an offering, anyone who practices

---

[242] Acts 13:8

[243] This is logical since by its very nature, magic is "apart from" God. Why would a believer need more supernatural input than what could be provided by God?

> divination or tells fortunes or interprets omens, or a sorcerer or a charmer or a medium or a necromancer or one who inquires of the dead, for whoever does these things is an abomination to the LORD. And because of these abominations the LORD your God is driving them out before you. (Deuteronomy 18:9-12).

The warning didn't end in the Promised Land. Paul tells us in 2 Timothy to avoid the "lying wonders" of those such as Jannes and Jambres, who were likely among the magicians in Pharaoh's court. Interestingly, these sorcerers were able to replicate some of the miracles performed by Moses…

> So Moses and Aaron went to Pharaoh and did just as the LORD commanded. Aaron cast down his staff before Pharaoh and his servants, and it became a serpent. Then Pharaoh summoned the wise men and the sorcerers, and they, the magicians of Egypt, also did the same by their secret arts. For each man cast down his staff, and they became serpents. But Aaron's staff swallowed up their staffs (Exodus 7:10-12).

There is no implication here that the sorcerers were employing illusions or sleight-of-hand. "By their secret arts" the Egyptian magicians (*hartemmim*), trained in occultism, could perform some of the same miracles as did Moses and Aaron—obviously empowered by fallen spirits. The thrust of this confrontation was not to convince Pharaoh of the supernatural but to prove that Yahweh God was superior to all of the Egyptians' spirit powers.

Modern Magic

There remains a dark side of magic today. Discussing occult magic, past and present, Unger believes…

> Such mighty demonstrations of demon power seem to occur periodically, like religious revivals, and are found today in Spiritism.[244]

Natural and personal magic are common today, despite the influence of Christianity. A large number of people, in a diversity of cultures, are involved with magic to varying degrees. Missionaries frequently deal with everything from folk magic to occult demonology. Tribal groups regularly use charms, curses, and talismans in attempts to manipulate reality. Shamanism is common in some traditional Native American groups but is equally welcome among modern spiritualists.

Even on a more innocent scale, I would venture to guess that the typical American doesn't step on sidewalk cracks or walk under ladders. And do you hesitate a moment when a black cat crosses, especially on Friday the 13th? You're probably not a full-fledged witch, but the point is, magic has *not* been "preached" out or "scienced" out of us in the 21st century.[245] Satan is more than happy to use even the slightest step into magic, making biblical warnings about magic as pertinent today as ever.

---

[244] Merrill F. Unger, *Biblical Demonology: A Study of Spiritual Forces at Work Today*, (Grand Rapids, MI: Kregel, 1994), 112.

[245] Be certain not to confuse entertainment "magic" with the darker sorts. Although modern illusionists often couch their acts in dark or demonic themes (an unfortunate but separate issue), most practitioners and observers realize that entertainment magic is simply well-rehearsed illusion and sleight-of-hand.

The Magic Christians

The intrinsic problem with magic is that it's an attempt to bypass God's way with a more "controllable" plan. We favor formulas in which we press a certain button and get a certain result. In its expected outcomes, magic is essentially a substitute for petitionary prayer. As with prayers of request, the magic practitioner's desire is a change in situation. It isn't surprising that Satan, the Master Counterfeiter, would attempt to capitalize on God's method (prayer) for allowing humans to supernaturally affect the physical realm.

But Christianity isn't based on a *recipe*. It's a *relationship* with a personal God who has the final say on a matter. Those who consider prayer a form of magic illustrate the problem. True scriptural prayer is a request nested in a relationship. It's never to be an incantation in which the correct phrase is uttered and the corresponding response occurs. We "let our requests be made known to God" and the outcome is totally up to him, yet always in our best interest. Jesus condemned the chanting mantras of the pagans and gave a new standard for relational prayer…

> When you pray, do not heap up empty phrases as the Gentiles do, for they think that they will be heard for their many words. Do not be like them, for your Father knows what you need before you ask him (Matthew 6:7-8).

It's disturbing to see a Christianized form of magic within modern churches. A subtle, sleight-of-hand can rewrite the sovereign response of the Most High into a formula that God's power *must* obey. This is often taught as the *power of spoken words*. Several Christian groups have the view that the "power" of God can be manipulated by using the correct words or procedures. In this belief, God is obliged to give a particular response when one performs the prescribed action: If I say "A + B", then "C" must happen. One very popular evangelist says…

> You can have anything you say!…the force of the spiritual world around us is controlled by the words you say. This force comes from inside you. (Paraphrased for anonymity)

Another in this same movement claims…

> If you are poor, change what you are saying and it will change what you have. Change your vocabulary. God is obligated to meet your needs because of His word.

This clearly substitutes a relationship with God with the "power" of God. It replaces God with the force of natural magic or uses him as just another entity who dispenses power by necessity as in personal magic. But God is not mocked. Christian-sounding "spells," even if they sound like prayers and may seem to work are as damnable as sorcery in God's sight…

> On that day many will say to me, "Lord, Lord, did we not prophesy in your name, and cast out demons in your name, and do many mighty works in your name?" And then will I declare to them, "I never knew you; depart from me, you workers of lawlessness" (Matthew 7:22-23).

Effectual Prayer

A counterfeit can only be such if there is a genuine article for it to mimic. The misuse of prayer should never eclipse the true gift God has given to believers. Volumes have been written about prayer, but the simple truth is offered by Pascal: "God instituted prayer in order to lend His creatures the dignity causality." I would add that this gift is ultimately for God's glory.

In a sense, prayer is the most imminent supernatural power possessed by believers. Without relinquishing his sovereignty, God has given us the ability to cooperate with him in altering the physical realm...

> The effectual fervent prayer of a righteous man availeth much (James 5:16, KJV).

Within God's will, the praying Christian can do that which the sorcerer can only dream. Could anything be as amazing?

If the Fates Allow

The terms *luck* and *fate* are so engrained in our culture that we often miss the jabs. Every time I open Google it temptingly probes how lucky I may be feeling. If luck is simply referring to a statistical probability of something, I could go with that, but there is usually a more supernatural implication. [246] If magic wasn't bad enough, the ideas of luck and fate imply that the mystical power is strewn randomly, with no rhyme or reason as to who gets it or when. Some try to sweeten the odds by avoiding black cats or carrying rabbit's feet, but the message is the same—sometimes you win, sometimes you lose. Clearly, luck and fate are foreign to biblical Christianity. Billy Graham puts it very bluntly…

---

[246] Although theologians might rightfully argue that there can't be even statistical "luck" if God is sovereign.

> If life is random or ruled by chance, then it doesn't really have any higher purpose. In fact, if you take it to its logical conclusion it means we are here by accident, and everything that happens to us is strictly accidental and has no meaning. Or we may decide we are ruled by fate and have no control over our destiny... But God tells us otherwise! God created us, and He put us here for a purpose. He loves us, and He cares about what happens to us![247]

Blessings and Curses

Blessings and curses are mentioned often in Scripture. As with many Biblical concepts, the words have become skewed in contemporary usage. In simplest terms, a blessing is a wish for good things to happen to someone. To wish bad things to happen would be a curse. We would all probably agree that to wish bad upon someone is at its least a sin of attitude. Beyond that, whether we wish goodness or badness on another is essentially a moot point. Wishing doesn't accomplish anything.

...Unless it's God doing the wishing... Only God has the power to actualize a "wish" as a real event.[248] Thus, we see God blessing and cursing in Scripture and his *will* becomes *event*. When God placed a curse on Cain, certain "bad things" were applied to the murderer. God's blessing of Abraham brought "good things."

The confusion, however, comes when we attribute blessings and curses to human agents. Sometimes, the word "blessing" in Scripture has the second meaning of a person's bestowal of certain rights on another. For example, Isaac's blessing of Jacob was more than just well-wishing. It was a validation of a certain inheritance or birthright. Also, a blessing

---

[247] Billy Graham, "Answers," www.billygraham.org, July 27, 2005.
[248] I would argue that if I attempted to actualize a wish by *doing* something to someone, it ceases to be a wish. This is not true with God.

may be prophetic or even a sort of prayer, requesting that God bestow a blessing.[249]

Curses are more black-and-white. First of all, they are forbidden by God's Word. Besides being a sinful attitude, curses are equated with spells and incantations. Therein lies the problem with modern cursing. (Not the use of profanity. That's a whole other problem.) From whom or what would one be invoking the curse? Is God being asked to do bad things to someone? That would be problematic, to say the least. The only alternative would be a form of *natural* or *personal* magic. Today, those who employ occult practices to pronounce curses seem to be invoking entities of some sort to empower their wish.

Recently, Wiccans across the country pronounced a *mass curse* on our president. Should the Secret Service be on high alert, or are curses just mythology? I don't believe there are magical "forces" that can energize malicious curses. Can fallen spirits carry out curses? If by that we mean "Can they cause bad things?" we know that is possible. But realistically, Satan and his team are already engaged in this as a continuing pursuit…

> Be sober-minded; be watchful. Your adversary the devil prowls around like a roaring lion, seeking someone to devour (1 Peter 5:8).

Evil spirits need no encouragement to be themselves. Nevertheless, if a believer is still worried that curses may be a threat, we have God's promise that…

> …neither death nor life, nor angels nor rulers, nor things present nor things to come, nor powers, [39] nor height nor depth, nor anything else in all creation, will

[249] Jacob's blessing of his sons seems to be bestowing inheritance, prophetic, *and* prayerful.

be able to separate us from the love of God in Christ Jesus our Lord (Romans 8:38).

This includes curses or any other malevolent intent from any creature. No evil will befall the believer apart from God's permissive, loving will.

Superstitions

Some forms of magic are so subtle as to zip right under our spiritual radar. The word "superstition" means to *stand over*, perhaps pointing to a rule that takes precedence over normal rules. Here again, the idea is that certain forces or entities are in control of events. As mentioned, many level-headed individuals who would never summon a spirit or use an incantation may go about life as usual on Friday the 13th, but with a bit more vigilance. We like the odds stacked in our favor "just in case."

I believe some superstitions are somewhat based on scientific reasoning, albeit flawed scientific reasoning. Remember, our minds are designed to find orderliness and purpose in randomness. That's a good trait. It allows the scientific method to work effectively in a universe bearing the stamp of purposefulness.

Assume you have no preconceived notions about black cats. Suppose a black cat crosses your path three times in a particular month. With each occurrence, you happen to recall that it rained each time within an hour. The scientific reasoning part of your brain kicks in—a pattern! *Does a black cat crossing my path = rain*? So far, this is still good science. You've formed a testable question.

There could be a real connection between the cat and the rain. What if the cat sensed the ominous clouds and headed home each time. The storm connected the rain and the cat's behavior. Or maybe a sadistic person always puts his rowdy cat out as a storm approaches.

With a potential superstition, however, the science stopped before this reasoning. Instead of testing the hypothesis by predicting and observing future crossings, you skip the step and make your question a principle: *Black cat crossing my path = rain*. But this isn't a tested hypothesis or principle. It's only a *pre*-superstition. Even if in the future a black cat crosses and it doesn't rain, you'll tend to ignore the contrary data. If it ever rains again after a black cat crosses, that incident will be far weightier than the contrary data. *A superstition is now established.*

Not only is superstitious reasoning bad science, it mocks Scripture and the nature of our Father God who is just as sovereign over ladders and black cats as over the movement of galaxies. Believers should steer clear of even the suggestion of powers beyond God's purview, and use the situation as an opportunity to grow in faith…

> Have nothing to do with irreverent, silly myths. Rather train yourself for godliness (1 Timothy 4:7).

18
Prophecy and Divination

"Man's interest in the future is legendary,
and many prophets—true and false—have tried to
satisfy that interest." Charles C. Ryrie[250]

"If a prophet or a dreamer of dreams arises among you and gives you a sign or a wonder, and the sign or wonder that he tells you comes to pass, and if he says, 'Let us go after other gods,' which you have not known, 'and let us serve them,' you shall not listen to the words of that prophet or that dreamer of dreams. For the LORD your God is testing you, to know whether you love the LORD your God with all your heart and with all your soul. You shall walk after the LORD your God and fear him and keep his commandments and obey his voice, and you shall serve him and hold fast to him.
But that prophet or that dreamer of dreams shall be put to death."
Deuteronomy 4:1-4

In the 1993 movie *Groundhog Day*, Phil Connors has the blessing (or the curse) of living the same day over and over. Through trial-and-error and observing each repetition in detail,

[250] Charles C. Ryrie, *A Survey of Bible Doctrine*, (Chicago: Moody Press, 1972), 159.

he gradually gains the knowledge to "perfect" that particular day of his life. How many times have you been asked: "If you could see the future, would you?"

Undoubtedly, Adam and Eve had it good. A paradise. No weeds. No mosquitoes. No telemarketers. No death. The Creator himself walked with them in the garden. Their *only* restriction was the tree of "the knowledge of good and evil." Was God withholding something they needed? Something good? This is purely speculation, but perhaps the Most High fully intended for them to *eventually* eat from this tree. After all, he isn't opposed to knowledge—at the proper time and in the proper context. Whatever the case, Adam and Eve, with the nudging of the Serpent, decided that *knowing* was better than *obeying*. They preempted God's plan and took matters into their own hands.

Despite what your junior high report card might show, we humans seem to have a deep-seated need to learn about things—especially when the information is secret or prohibited. Of course, knowledge isn't a bad thing. Proverbs and many other scriptures encourage us to seek and to treasure knowledge, understanding, and wisdom. Here again, the problem lies not in the innate desire, but the mode and extent of our learning.

Our fallen nature often leads us to pursue knowledge before its time, beyond that which is appropriate, and outside of God's method. As children of Adam and Eve, we impatiently grope for ways to bypass God's strategy for revelation. Much of human history deals with this persistent grabbing for forbidden fruit.

Prophecy vs. Divination

The most obvious bank of "secret" knowledge is the future. It looms before us like a thick fog. Information comes into sight one nanosecond at a time as we press forward. Occasionally, at his discretion, God blows away the mist and allows a distant panorama. A significant portion of the Bible is this extended view called *prophecy*. The broader definition of prophecy, however, includes God's declaration of *any* truth through a human agent.

In Scripture, God often spoke through prophets, his designated, earthly messengers. Biblical prophets spoke and wrote via the direct inspiration of the Spirit of God. By faith, Christians accept their prophecies as the inspired Word of God. The prophets' notarization: "Thus says God."

The term *divination* is used in Scripture to designate forms of knowledge-seeking that are outside of God's will. Unger's description shows its similarity to biblical prophecy…

> The basic presupposition underlying all methods of divination is that certain superhuman spiritual beings exist, are approachable by man, possess knowledge which man does not have, and are willing, upon certain conditions known to diviners, to communicate this information to man.[251]

But these general assumptions are merely superficial similarities. In contrast to biblical prophets, the pagan *mantis* or *soothsayer* typically relied on chants, drugs, motions, and altered states of consciousness to divine the message. It was common for the mantis to eat organs of a sacrificed animal in order to establish a *channel* for communication.[252]

As with magic, there are two analogs of divination. In *artificial divination* (also called *augury*), the mantis "reads" a sign or *omen* to gain special, hidden knowledge. An omen could

---

[251] Merrill F. Unger, *Biblical Demonology: A Study of Spiritual Forces at Work Today*, (Grand Rapids, MI: Kregel, 1994), 120.

[252] Similar practices have continued to the present. Besides pagan groups and nature religions, New Agers and others use mind-altering drugs, not simply for recreation but spiritual enlightenment. A more refined upgrade of LSD, DMT, is touted as a link to "machinery elves" and other entities willing to provide "forbidden" knowledge.

have taken a variety of forms such as tea leaves, Tarot cards, zodiacs, or even livers. The assumption was that a supernatural agent manipulated the omen, infusing it with the desired knowledge. For example, it was standard practice for some pagans to read the liver of the sacrifice as an omen. Livers were believed to be the container of the soul; thus, the gods or demons could "connect" with the liver to relay information. Note that the message originated from a supernatural being, but the mantis had to be skilled in the art of extracting the message.

Inspirational divination posits that a supernatural being "inhabits" the soothsayer or "endues" him with the actual message, perhaps *verbatim*.[253] In other words, the diviner may possess no special skills. He is merely the conduit through which another being speaks or writes. This is sometimes referred to as "mechanical" or "automatic" communication, meaning that the mantis was in a trance-like state, unaware of the event.

Nostradamus

One of the most popular prophets of the western world, Michel de Nostredame, or Nostradamus, lived in 16th-century France. He was born into a Jewish family who had converted to Roman Catholicism. After working several years as a physician and apothecary, Nostradamus began experimenting with the occult practices of astrology, charms, and necromancy. He is best known for the publication of *Les Prophéties* (*The Prophecies*), a large collection of enigmatic predictions of the future world. Fearing the Inquisition, Nostradamus always denied being a true prophet: "Although, my son, I have used the

---

[253] Some claim that this is the same as Biblical inspiration, although the mainstream view is that the writers of Scripture were inspired through their own unique personalities—the result being the exact intent of God.

word *prophet*, I would not attribute to myself a title of such lofty sublimity."[254] He subsequently obscured the meanings of the prophecies by formatting them as quatrains (four-line poems).

Les Prophéties has never been out of print since its first publication, and literally millions of would-be interpreters have contributed their two cents. The resulting babel is an excellent study into what can happen when ambiguous predictions are pressed into the mold of pop culture. The first important element in creating a convincing prophecy is to speak vaguely. For example, one of Nostradamus's most notable prophecies has been linked to Adolf Hitler…

> Beasts ferocious with hunger will cross the rivers,
> The greater part of the battlefield will be against Hister.
> Into a cage of iron will the great one be drawn,
> When the child of Germany observes nothing.

With our hindsight knowledge of WWII events, some of these phrases might seem applicable. But looking at history in general, they could relate just as well to any number of events. Is the reference to Germany unusual? It might be—if Nostradamus lived in Australia. Germany, however, was just across the border. The other proper noun in the quatrain might seem equally conspicuous, but *Hister* is a region around the Danube and isn't etymologically related to *Hitler* in any way. Realistically, if the words *Germany* and *Hister* didn't appear in such proximity, it's doubtful that anyone would have ever made a connection.

Another strategy for succeeding as a false prophet is to make so many predictions that some might actually hit close to the target. (Even a stupid squirrel finds a nut every once in a while.) Nostradamus wrote thousands of prophecies. Over a five-hundred-year period, we would expect that many of the broad predictions would mirror certain events. Nevertheless,

---

[254] Nostradamus, *Preface to César*, 1555.

Nostradamus groupies still hammer away, trying to fit the prophetic pegs into historical holes.

Finally, a good false prophet needs a receptive and undiscerning audience. While the Twin Towers were still burning on 9/11, this Nostradamus quatrain appeared on the Internet…

> In the City of God there will be a great thunder, Two brothers torn apart by Chaos, While the fortress endures, the great leader will succumb, The third big war will begin when the big city is burning.

Sound convincing? After viral reaffirmation of Nostradamus, it was learned that a university student had written this particular quatrain in 1997 as an illustration of how easy it is to write an authentic Nostradamus-style prophecy. (Wait… isn't everything on the Internet true?)

The Modern Mantis

What about contemporary prophets who seem to hit it "right on the nose?" In 1956, self-proclaimed prophet Jeane Dixon prophesied that the winner of the 1960 presidential election would be assassinated. What were the chances that such a specific prophecy would come to pass with the death of JFK? Does this rise to the level of Old Testament accuracy or even a miracle? Presidential assassinations in the U.S. are somewhat rare, but once again, context reveals much. That prediction does seem remarkable until we look at her others. Dixon also predicted that Kennedy would *not* win the 1960 election. Furthermore, she erroneously predicted that the assassination would occur in the president's *second* term. Would she pass the Deuteronomy test?...

> But the prophet who presumes to speak a word in my name that I have not commanded him to speak, or who speaks in the name of other gods, that same prophet shall die. And if you say in your heart, "How may we know the word that the LORD has not spoken?"— when a prophet speaks in the name of the LORD, if the word does not come to pass or come true, that is a word that the LORD has not spoken (Deuteronomy 18: 20-22).

Modern divination is yet another example of the paranormal paradox within the Church. It's disheartening that many who claim a Christian worldview also make allowances for those "innocent" forms of divination. "Social dabbling" into the occult is currently increasing, especially among young people.[255] Is this experimentation simply hobby-sorcery or is it more perilous?

For as long as there have been stars to ponder, humans have attempted to tie their motions to life here below. *Astrology*, possibly the oldest "science," has been practiced continuously and is as popular as ever.[256] Somewhere in prehistory, skywatchers noted the consistent, predictable motions of celestial bodies. They learned that the star patterns followed an imaginary line (the *ecliptic*, actually derived from the rotation of the earth) through the night sky. The sun tracked through twelve different star groupings along the line in a single year (the zodiac constellations).

---

[255] Jennifer LeClaire Ministries, "Ouija Boards Are Growing in Popularity Among Christians," www.christianpost.com, November 30, 2017.

[256] We now differentiate between astrology (the pseudoscience) and astronomy (the scientific study of celestial objects). The lines were blurred until the age of modern science. To varying degrees, early astrologers studied both the science and the "sorcery" of the heavens. Although steeped in the occult, many astrologers calculated celestial motions quite accurately, making important contributions to science.

Many were in awe of how "the heavens declare the glory of God." Others read into the intricacies a source of power and secret knowledge. Today, hundreds of millions of people base their activities to some degree on personal horoscopes. Modern astrologers have the benefit of computers to calculate and disseminate divinations in nanoseconds.

Despite new technology, the accuracy of the predictions remains scientifically and practically abysmal. Apart from scattered sheer coincidences, horoscopes depend on the fact that generalities will always seem to fit in *some* way. For example, I looked up my own horoscope for today…

> *[Date, sign] News could come your way, perhaps through local magazines or newspapers, about possible renovations occurring in your community. These changes could be rather controversial, [sign], so don't be surprised if you hear a lot of gossip and a number of opinions, both pro and con, on the matter. You might want to delve more deeply into the facts before forming an opinion of your own. This could prove quite enlightening!*[257]

I'm a teacher. This could virtually describe every day of my life! For the Christian, horoscopes are essentially a no-win game. If the message is wrong or unproductive, you lose. If the message seems to help you in some way, congratulations! You've just employed a fallen spirit to circumvent God's perfect, loving will for you! But aside from merely pragmatic reasons to avoid astrology, Scripture is clear about the practice in general…

---

[257] www.horoscope.com; On a link to another site, I could get a live, full reading with a professional astrologer (with credit card).

> Let them stand forth and save you, those who divide the heavens, who gaze at the stars, who at the new moons make known what shall come upon you. Behold, they are like stubble; the fire consumes them; they cannot deliver themselves from the power of the flame (Isaiah 47:13-14).

A Christianized version of the zodiac has been circulating since a book by Frances Rolleston in the 1800s.[258] The premise is that God posted the timeline of his redemptive work in the zodiac. Rolleston's scriptural basis was derived from Psalm 19…

> The heavens declare the glory of God, and the sky above proclaims his handiwork. Day to day pours out speech, and night to night reveals knowledge. There is no speech, nor are there words, whose voice is not heard. Their line goes out through all the earth, and their words to the end of the world (Psalm 19:1-4).

Rolleston interpreted the "information" metaphors in the passage as a prophecy of Christ. Many scholars agree that the translated term "line" does refer to the ecliptic line, and the language seems to imply the display of a "message." Several prominent Christian leaders agree with this *gospel in the stars* (GIS) idea.[259] In this version, Leo the Lion represents Christ's Lion of Judah title, and Virgo is the Virgin Mary. Scorpio is Satan, Aries the Ram represents sacrifice, and so on. I enjoy stargazing, and while doing so I often mull over this galactic redemption story. I admit it offers good talking points and it's an interesting teaching tool. I'm not as certain, however, about

---

[258] Frances Rolleston, *Mazzaroth, or the Constellations,* (London: Rivingtons, Waterloo Place, 1862).
[259] Charles Strohmer, "Is There A Christian Zodiac, A Gospel in the Stars?" Article ID: DG240, www.equip.org, 22 Apr 2009.

whether the intent of Psalm 19 is to lay out a new component of natural revelation.

Modern sorcery isn't limited to the celestial realm. Since the movie *Ouija: Origin of Evil*, sales of the "toy" have exploded.[260] (I wonder if anyone saw that coming?) The Ouija board was originally introduced by Parker Brothers as a "talking board" parlor game in 1890. Occult connections quickly emerged as spiritualists had long been using similar *spirit boards* as necromancy tools to contact the dead. Ouija messages are considered a type of automatic writing in which a spirit entity controls the movement of the planchette (wooden cursor) as it moves to various letters.

The Internet facilitates a seemingly endless variety of sorcery games. Charlie Charlie (*Juego de la Lapicera*) has become one of the most popular among teens worldwide. Divination is cunningly marketed to unwary Christians with *angel boards*, allowing messages from "friendly" Ouija spirits. An "angelic" version of tarot cards is also available. Many view even the frivolous entertainment use of horoscopes, Ouija boards, and similar pursuits as introductions to more sinister occult practices…

> Ultimately, it's never a simple and harmless thing to use any method to contact the dead and/or the spirit realm, it is a doorway into the occult.[261]

We can't accuse the modern spiritualist of being behind the times. For the tech-savvy there are a number of electronic options for contacting spirits. *Electronic voice phenomena* (EVP) can be detected and recorded by gadgets made expressly for that purpose. The cheaper models are simply static

---

[260] Neil Tweedie, "Sales of Ouija Boards Up 300%," www.dailymail.co.uk, 30 Nov 2014.

[261] Matt Slick, "Can a Christian Use a Ouija Board?" www.carm.org, Accessed 10 Feb 2020.

generators and rely on audio pareidolia, the tendency of the mind to perceive sense in nonsense. Users claim to hear whispered words or phrases emanating from the static. High-end EVP devices have a digital bank of words spoken by a voice simulator. Words are said to be produced as a spirit "wills" them. These high-tech necromancing tools come with the same hazards as Ouija boards. Most of the messaging is simply randomness, but if a fallen spirit wished to deceive digitally, how nice to have an open channel and an anxious listener. Billy Graham expounds the practical effects of occult dabbling, in general, pointing to the relational aspect…

> Have you ever met anyone who was involved with the occult who truly knew God as their friend and loved Him with all their heart and mind and soul and strength?... If you're honest, you know the answer… The reason is simple: The occult—whatever form it takes—will lead you away from God, not toward Him.[262]

Paul explains in detail the freedom we have as Christians, in this case, the fact that evil spirits have no claim on us. Many believers have engaged in occult-based games as entertainment with no serious consequences. Even though strong Christians *could*, should they? Paul goes on to say that with any freedom comes responsibility (1 Corinthians 8). Do our actions affect "weaker" brothers and sisters? Believers should keep this principle in mind in all paranormal dabbling.

---

[262] Michael Gryboski, "Billy Graham Answers: Are Ouija Boards Harmless Fun or Satanic?" www.christianpost.com, November 03, 2016.

We Three Kings of Orient Are What?

Likely you've heard the sad news. Our Christmas plays have lots of mistakes. There probably weren't three, and they really weren't kings…

The New Testament gives few details about the wise men (*magi*). The title itself informs us that these men were Gentile magicians, highly trained in sorcery and other academics. Originally centered in the Medo-Persian/Babylonian regions, some magi had spread throughout the Roman Empire at this time due to persecution. Many Jewish people were self-proclaimed magi, such as Simon the *Magus* in Acts 8.

These magi were likely legit, however, as we see them coming "from the East." We know something of their training through Daniel's encounter. Magi were held in high esteem, being educated not just in sorcery, but science, math, and the philosophical wisdom of the day. Because of their accumulated knowledge, one of their roles was the selection of kings and other leaders. These particular magi seemed to be quite familiar with the Jews and their God. It seems reasonable that Daniel might have played a part in this centuries earlier…

> Because of Daniel's high position and great respect among them, it seems certain that the magi learned much from that prophet about the one true God, the God of Israel, and about His will and plans for His people through the coming glorious King.[263]

Scripture indicates that a star informed these magi of Christ's birth. Despite tradition, there is much ambiguity in this scripture. We simply don't have information regarding the nature of this sign. There has been speculation about eclipses,

---

[263] John MacArthur, "How Did the Magi Know About Jesus?" www.gty.org, 27 Jun 2019.

comets, supernovas, UFOs, and every imaginable "star." All we know is that God allowed them to discern a clear indication of Jesus' birth. The important aspect is that the magi responded.

So, did God communicate his son's birth through pagan sorcery? Without more information, this is the point at which we depend on God's unchanging, righteous nature to fill in the blanks. Having condemned all types of divination, would he use these practices? I tend to believe he wouldn't. But even if these "pagans" were bent on using stars in an ungodly way, perhaps God intervened and sent them true light.

Furthermore, nothing in the account implies a justification of astrology. An often overlooked aspect of the nativity is its foreshadowing of the gospel's broader application: In the short-term, Herod and the Jewish nation would reject the Savior while the Gentile nations come to him.[264] It's clear by the magis' words and actions that they understood the Hebrew prophecies concerning the messiah quite well, perhaps better than the Jews they encountered. I would like to believe that God in his mercy, revealed Divine truth to the wise men and that they continued to respond to the light they were given.

Prophets in Church

The role of prophet in today's Church is controversial. Some denominations teach that God continues to deliver new truths via selected individuals. Indeed, the act of prophesying is an expected, regular part of many worship services. The consensus of most who hold to a literal interpretation of the Bible, though, is that the role of prophet ended as the completed copies of the New Testament began circulating.

Prophets in the 1st-century Church were necessary because new revelations concerning Christ might not have been available to particular local churches. Their messages were not

---

[264] Matthew 28:19; Isaiah 60.

so much *fore*telling the future, but *forth*telling truths about Christ.

These prophetic teachings might have been new revelations to the specific congregations, but they were subject to the strictest scrutiny. Anything spoken by a prophet must not disagree with Old Testament teachings. Also, all that a prophet spoke had to mesh with the teachings of the Apostles. God was in a sense, leading them to "fill in the blanks" of the gospel being presented (through the guidance of the Holy Spirit). With all teachings tested against these criteria, the application of the truths would also be fruitful and edifying to the church. After the local churches had received the written Word of God, this interim mode of prophecy was replaced by the gospel of Christ (New Testament)...

> Long ago, at many times and in many ways, God spoke to our fathers by the prophets, but in these last days he has spoken to us by his Son, whom he appointed the heir of all things, through whom also he created the world (Hebrews 1:1-2)

John MacArthur further explains...

> In the early Church, the Apostles were the evangelists and the prophets were like the original teaching pastors, and that's why they are replaced [today] by evangelists and teaching pastors.[265]

From a practical standpoint, one who presumes to be a prophet in today's Church is either redundant or cursed. All that God has intended for us to know is already in his Word. If

[265] John MacArthur, "John MacArthur on Modern Prophets," www.gty.org, 3 Sep 2013.

something new is being added, the specific warning in Revelation seems to apply to inspired Scripture in general…

> I warn everyone who hears the words of the prophecy of this book: if anyone adds to them, God will add to him the plagues described in this book, and if anyone takes away from the words of the book of this prophecy, God will take away his share in the tree of life and in the holy city, which are described in this book (Revelation 22:18-19).

Deception in Prophecy

In a complete overview of prophecy, we should consider whether the Ephesians Hypothesis might have relevance. Concerning modern prophets, we've seen that coincidence and lack of discernment among recipients can explain most of the isolated "successes." The track records of contemporary soothsayers should be enough to discredit all of them. Still, advocates cling to those predictions which seem supernaturally inspired.

There have been prophets who so adulterate the Scripture that we could assume Satanic influence. Certainly, the prophecies of cult leaders such as David Coresh, Jim Jones, and Marshall Applewhite were so compelling that demonic assistance wouldn't be out of the question. Even "miraculous" individual prophecies associated with horoscopes and psychic readings could logically be attributed to fallen spirits. As previously noted, demons aren't omniscient and they probably can't foretell the future, but they can likely see a broader perspective, and to some extent, they might even manipulate the outcome of a prophecy.

There is little doubt that Satan and his minions have the ability and the motivation to affect the minds of both the so-called prophet and the hearer, as well as to manipulate the accompanying situations. Any or all of these could be used in leading both nonbelievers and undiscerning believers away from truth. Is it any wonder the Old Testament gives such strict warnings against false prophets?

But the standard was no lower in the New Testament. Paul and Silas encountered a slave girl possessed by "a spirit by which she predicted the future."[266] Instead of lecturing on the possible dangers of occult practices, they go right to the root of the problem: "[Paul] turned around and said to the spirit, 'In the name of Jesus Christ I command you to come out of her!'" Obviously, not every modern-day soothsayer is demon-possessed, however, Scripture affirms the power source for all occult phenomena. There is no place for it in Christian practice or worldview.

Dreams

I'm skeptical about the significance of dreams in general. Yes, I know psychologists have studied them at length and have long lists of symbols and archetypes. Maybe I'm wrong, but my dreams seem to be random, dumb things that are more embarrassing than informative. (If you are a psychologist reading this, I don't want to know whether my disregard for dreams has some symbolic meaning.) The surreal nature of dreams in general and their almost nightly occurrence make them likely avenues for misinterpretation, deception, or confusion. Should God choose to speak through a dream today, it would be the exception rather than the rule, and it would conform to the precedent we have in the Bible.

There are instances in Scripture in which God used dreams and visions to communicate messages. As with other supernatural interventions, the dreams were rare, purposeful,

---

[266] Acts 16:16

and contextually relevant. Peter had a vision of a sheet with unclean animals, illustrating that God would now be dealing with Gentiles. John's vision was used to teach him the complicated events to come. God could have communicated differently, but these concepts were likely more effective in visual format.

Daniel and Joseph were given the ability to interpret dreams. In their cases, dream-reading was a badge of authority leading to new opportunities for service. God's Word confirms that he is much more likely to communicate through the acts of his son (the living Word) and the truths about him recorded in the written Word. Should one be convinced that God has spoken through a dream or vision, it would be prudent to use Tozer's guidelines for discernment stated earlier.

The Biblical Standard

The masses cheer when a modern soothsayer offers a vague, general prophecy that "sort of" comes to pass. Biblical criteria for prophetic accuracy were considerably higher. All the prophecies in Scripture are fulfilled precisely with no hint of error, but some draw our attention more than others. Predictions such as the mathematical forecast made in Daniel illustrate the mark to which a true prophecy must align…

> Seventy weeks are decreed about your people and your holy city, to finish the transgression, to put an end to sin, and to atone for iniquity, to bring in everlasting righteousness, to seal both vision and prophet, and to anoint a most holy place (Daniel 9:24).

The Hebrew word translated "week" is simply the word "seven." Scholars agree that this passage is referring to seventy *sevens* of years (490 years). But this 70 "week" period of God's dealing with Israel would have two parts…

> Know therefore and understand that from the going out of the word to restore and build Jerusalem to the coming of an anointed one, a prince, there shall be seven weeks. Then for sixty-two weeks it shall be built again (Daniel 9:25).

If you followed the math, there will be 69 "weeks of years" (7 + 62) between the decree to rebuild the temple and the coming of Messiah. Using the Jewish calendar, that comes to 173,880 days.[267] We know that Artaxerxes Longimanus gave the rebuild order on 14 March 445 BC. What was the actual day of Jesus's arrival as Messiah? The key is in the word "prince" (which is synonymous in the original with "king"). Jesus fulfilled the prophecy of Messiah's official "arrival" given by Zechariah…

> Rejoice greatly, O daughter of Zion! Shout aloud, O daughter of Jerusalem! Behold, your king is coming to you; righteous and having salvation is he, humble and mounted on a donkey, on a colt, the foal of a donkey (Zechariah 9:9).

Jesus formally proclaimed his role as he rode into Jerusalem on 6 April 32 AD. Although the Pharisees urged him to tell the crowds he was not Messiah, Christ validated the people's declaration: "I tell you, if these were silent, the very

[267] Our calendar year is based on the 365-day solar calendar. The Jews and others of the time used a 360-day lunar calendar year.

stones would cry out."[268] The event occurred *exactly* 173,880 days after Artaxerxes' order.[269]

Some skeptics argue that Jesus had control over the fulfillment of this particular prophecy. I suppose he did decide to ride a colt into Jerusalem on that day, but it's certainly difficult to explain the "lucky" series of events leading up to this point in his life. Skeptical reasoning of this sort seems to take more faith than belief in the supernatural. This complex prediction obviously sets a very high standard for prophecy.

---

[268] Luke 19:40

[269] Premillennialists place the remaining 7 weeks (years) as the Tribulation period. At that time (after the removal of the Church in the rapture), God will again focus on the Jewish people.

19
Parapsychology and Other Mind Games

> *Feb 13, 2020 - In a last-ditch effort to catch a fresh lead on the Jacqueline Brown case, the Lavana County Sheriff's Office followed the advice of deputies. Local psychic Dana Cusnack was contacted by Sheriff Dunn earlier this week and within twenty-four hours Brown's remains were located at a secluded area north of Twin Lakes. "Naturally, I was somewhat skeptical," Dunn stated in a news release last night. "But I have to admit, Cusnack's 'readings' were instrumental in the results."*

How many times has this scenario been played out in law police departments across America? Well, actually—few if any. This is fake news. I just made it up. Judging by TV cop shows, though, law enforcement would screech to a halt without the aid of psychics visualizing crime scenes and "postcognating" the suspects. The truth is, hardly any crime-solvers use the services of psychics.[270] Law enforcement does deal with psychics often, though. When major cases go public, investigators typically must wade through piles of incoming erroneous psychic "tips," simply because all tips must be checked. (Why don't the

[270] Erica Goode, "Police Rarely Turn to Psychics, Texas Report Aside," www.nytimes.com, 8 Jun 2011.

psychics know about the cases before they go public?) And despite what we see on TV, track records of real-life psychics are dismal.

Parapsychology is the general term for several alleged psychic phenomena. The name does seem to lend a measure of credibility to its practice. In the 20th century, some scientists began research on these alleged mental powers, with the assumption that a little-known physical force known as *Psi* was the source. Early proponents reported solid evidence for this paranormal power, but critics were quick to note the nuisance variables associated with human subjects and the unreliability of mental measurements.

It's interesting that in the paranoia of the Cold War, both America and the Soviet Union undertook serious research into psychic ability as a weapon. The investigations of the CIA's Project Stargate were supposedly abandoned in 1995, due to lack of any practical value in intelligence gathering.

Déjà Who?

Human bodies are, indeed, "fearfully and wonderfully made." I can't think of anything in God's physical creation that surpasses the intricacy of the brain. (I'm amazed that I can even think about my brain.) At times, certain brain functions seem to verge on the supernatural. Most of us have experienced the sensation of *déjà vu,* the feeling you have lived through a situation before. Some have tried to link this to precognition, but its association with physical brain events seems to indicate that it's merely an organic phenomenon.

Even people who don't consider themselves psychics might claim to receive *premonitions* (feelings or visions about the future) that eventually come true. This is a common occurrence and it's difficult to imagine there isn't something supernatural going on. Much of the time, however, these can be attributed to the same statistical flukes that cause superstitions and false miracles. It's part of our human psyche to connect events so that they make sense. (This is much like how a worldview works.) We wouldn't function very well without this cause-and-effect, logical thinking. But that same trait can sometimes work overtime and see connections that that are simply the result of random chance.

For example, the thought enters your mind of how good a plate of lasagna would be. That night, you go home and… Wow! Your spouse has made lasagna! How can that not be supernatural? The simple answer is that you have hundreds of thousands of thoughts per hour. In a single day, imagine how many of those predictive type thoughts *didn't* come to pass. The one random thought out of millions that just happens to occasionally come true is instantly connected to the real event and a "premonition" is born. We tend to notice connections over non-connections.

I'm not dismissing the possibility of something supernatural. God could use premonitions in some way, but I believe we should judge them as we do other miracles. Is it likely that God *would* use a miracle or premonition rather than some more ordinary means? *How does he usually reveal truth?*

And even if one is convinced that a premonition is supernatural, is it necessarily from God? Would a fallen spirit have any reason to lead our thoughts in a particular way? Would that deceiving spirit gain from a person's belief that a premonition is real? As always, we should "test the spirits."

Extrasensory Deception

Parapsychology deals with more than simply the "forward" look. The term *clairvoyance*, meaning "to see clearly," refers to any direction of the mind's paranormal gaze. One clairvoyant tool is *psychometry*, in which a psychic touches an object such as a pen and claims to be able to discern who has written with it.

Precognition, or *prescience*, is the ability to "see" future events—another flavor of divination or prophecy. The phenomenon, in general, is more personal than prophecy, usually informing one person about a specific future event. The seer might claim to see in several ways. Traditionally, the practitioner gets mental pictures regarding a certain scenario. The images typically come during a trance state or in a dream.

Postcognition is also an interesting concept. This ability to "see" events that happened in the past is like history class with a dash of paranormal. The practitioners claim to be able to discern the details of a past event, usually as a vision. Precognitions are typically of obscure events such as secret crimes or a day in the life of Abe Lincoln. Parapsychologists attempt to add a scientific base by hypothesizing a type of energy field that holds a record of the past like a paranormal hard drive.

Jesus seemed to use postcognition when he told both Andrew and the Samaritan woman what they had been doing. Is there a place for this phenomenon today? It seems to be similar to the spiritual *gift of knowledge*. Modern charismatics claim this gift still operates in the ability of some believers to "know" what is happening to another person, not in direct communication, usually for the purpose of prayer. Since it mostly in the present, the term *extrasensory perception* (ESP) or *second-sight* might be more accurate than *postcognition*. Some TV evangelists tout evidence that word-of-knowledge healings are legitimate. It's difficult to gather hard data since the individual accounts are usually quite general. Also, with a large enough audience, any pronouncement will likely apply to someone. Could God communicate through words of knowledge? Would he?

On another level, however, it may be difficult to discern some supernatural words of knowledge. Isn't every prodding of the Holy Spirit a supernatural word of knowledge? How often in our moment-by-moment walk does the Spirit give us a fresh dose of wisdom or understanding? We are promised these interactions. Paul prayed specifically for the Spirit to impart such needs…

> …that the God of our Lord Jesus Christ, the Father of glory, may give you the Spirit of wisdom and of revelation in the knowledge of him, having the eyes of your hearts enlightened, that you may know what is the hope to which he has called you, what are the riches of his glorious inheritance in the saints (Ephesians 1:17-18).

How do we know that it's God's Spirit communicating? Unlike random paranormal experiences, God's working is within a context. Meyers assures us that we have a three-fold witness for a true work of the Spirit…

> God's impressions within and his word without are always corroborated by his providence around, and we should quietly wait until these three focus into one point.[271]

Postcognition is often submitted as evidence for reincarnation. There are several eerie cases in which a person remembers places and incidents that aren't part of their known experiences. Although anecdotal, the subjects are able to show an uncanny familiarity with someone from the past, apart from recorded historical records. One well-known case involves Barbro Karlen who believed herself to be the reincarnation of Anne Frank.[272] At an early age, Barbro began remembering

---

[271] F.B. Meyer, *The Secret of Guidance,* (Chicago: Moody Publishers, 1896, updated 2010), 28.
[272] Barbro Karlen, *And the Wolves Howled: Fragments of Two Lifetimes*, (Clairview Books, 2000).

aspects of Frank's life which presumably she could not have known. Upon visiting Frank's former home, Barbro was able to locate and identify various features.

This account could be multiplied by hundreds. Similarly, subjects are often questioned by experts who verify the "remembered" information about a past life. Is this proof of reincarnation? On the surface, there does seem to be a supernatural element. Without firsthand knowledge of these scenarios, I could only offer opinions based on a biblical worldview, which rejects the possibility of reincarnation. What would be an alternative?

Again, this is speculation, but I believe it to be consistent with the Ephesians Hypothesis. We established earlier that fallen spirits are able to plant and lead thoughts. I see no reason why these spirits would not know historical details better than our recorded history (they were there, ostensibly). Would they be able to relay obscure yet accurate historical data? Would the phenomenon seem to have the same outcome as a reincarnation episode? Would fallen spirits have anything to gain by "faking" evidence for an unscriptural phenomenon?

Hypnosis

The dictionary definition of *hypnosis* is "a trancelike state that resembles sleep but is induced by a person whose suggestions are readily accepted by the subject."[273] Medical practitioners expand the characterization…

> …hypnotherapy or hypnotic suggestion is a trance-like state in which you have heightened focus and concentration… When you're under hypnosis, you

[273] "Hypnosis." *Merriam-Webster.com Dictionary*, www.merriam-webster.com, Accessed 23 Mar 2020.

> usually feel calm and relaxed, and are more open to suggestions.[274]

And with that, the agreement ends…

Trance-like states have been common since antiquity, but the modern form was popularized by occult-dabbler Franz Mesmer around 1780. Mesmer believed hypnosis (mesmerism) was enabled by some "animal magnetism" flowing from the hypnotist into the subject. Scientists quickly redefined it as a purely physiological phenomenon.

From a psychological standpoint, hypnosis leads a mind to a specific area of concentration without mental or physical distractions. It has been used in pain management, memory retrieval, behavioral therapy, relaxation techniques, and entertainment. The phenomenon isn't entirely understood, and its clinical efficacy is debated. There is also disagreement among Christians as to its usefulness and safety. Hypnosis is regularly used by some Christian psychologists as a tool for a range of therapies…

> We realize that hypnosis has acquired a bad name in the minds of many Christians. For our part, we see no reason to regard it as "evil" or "dangerous" in and of itself. In actuality, there is little or no basis for most of the popular fears associated with this technique. It's not true, for instance, that a hypnotist can gain control or mastery of another person's will.[275]

Others warn against any use of hypnosis, based on its association with other ideologies…

---

[274] Mayo Clinic, "Hypnosis," www.mayoclinic.org, Accessed 23 Mar 2020.
[275] Focus on the Family, "Hypnosis as a Clinical Tool," www.focusonthefamily.com, Accessed 23 Mar 2020.

> Altered states of consciousness and other mystical practices open doors in the spiritual realm. Once a door is opened, a person becomes vulnerable to any spirits which come through that door.[276]

Secular advocates of hypnotism also warn against its misuse. The Mayo Clinic recognizes side effects such as headache, anxiety, distress, and the creation of false memories.[277] Experts agree that hypnosis should be used only by trained professionals. The consensus among Christians seems to be that unless hypnosis is clearly useful and practical in a particular instance, it should not be used. It's also clear that the worldview of the hypnotist is a significant factor in the results obtained. Connections to occult and New Age activity should be avoided…

> …hypnosis, like any other therapeutic technique, should be rejected out of hand when it takes on questionable spiritual overtones – for example when it becomes occultic in nature or is clearly connected with the tenets of eastern religion or New Age philosophy.[278]

Telekinesis

The concept of "mind-over-matter" is quite ancient. Eastern mystics have long taught that the conscious self can alter physical reality. If I had my choice of superpowers, it would be a hard decision between telekinesis and invisibility (although flying might beat both). I have to admit, not having

---

[276] Dan Delzell, "Hypnosis and Yoga Open Hidden Spiritual Doors," www.christianpost.com, 28 May 2012.
[277] Mayo Clinic, "Hypnosis," www.mayoclinic.org, Accessed 23 Mar 2020.
[278] Focus on the Family, "Hypnosis as a Clinical Tool," www.focusonthefamily.com, Accessed 23 Mar 2020.

to get up to retrieve the TV remote would be very practical. One of the many interpretations of quantum physics seems to evoke mind-over-matter, but most physicists find too many serious flaws in telekinesis to consider it an option.

Some psychics claim that telekinesis (or *psychokinesis*) could be a spirit causing objects to levitate, move, or be thrown across the room. That would fit the concept of an alleged poltergeist. Most proponents of telekinetic ability believe the power is truly paranormal and has as its explanation some unknown force.

Uri Geller kept us all amazed and entertained with his telekinetic spoon-bending, as have many other magicians. All have been shown to be illusionists' tricks. Physicists claim that it would be easy to scientifically test the phenomenon. Surely ultrasensitive probes could detect the motions caused by a mental command. So far, no one has pursued this scientifically. Also, given the similarity to poltergeist phenomena, I see no reason why demonic entities couldn't "cooperate" in a telekinetic event.

If You Could Read My Mind

Deanna Troi (*Star Trek's* Betazoid counselor) is an *empath*, possessing the ability to sense another individual's emotional state. Most of the crew avoided poker games with her. It would be bad enough for someone to know my feelings, much less my exact thoughts. If I encountered a true mind reader, I think my first thought would be "If you can really read my mind, blink twice." If they blinked, I would run as fast I could to a safe "out-of-range" distance. We love the safety of our own private thoughts.

Sideshow psychics tease audiences with some amazing feats of mind-reading. But here's the deal: So do stage magicians. Illusionists usually don't claim psychic abilities, rather they will reluctantly admit it's all a clever trick. The Americans, Russians, and Nazis have all experimented with person-to-person mind reading as a weapon. After wasting much time and money, the conclusion was that mind-reading isn't possible.

However, God isn't so limited. My Creator, in perfect love and omniscience, knows my thoughts—even better than I do! (Fortunately, he is merciful and forgiving.) Several passages affirm this truth…

> Sheol and Abaddon lie open before the LORD; how much more the hearts of the children of man! (Proverbs 15:11).

> O LORD of hosts, who tests the righteous, who sees the heart and the mind, (Jeremiah 20:12).

Most theologians believe that Jesus purposely didn't draw on some of his attributes while in his earthly body.[279] There were some things that may not have been known to the man Jesus since Scripture says he "grew in wisdom and stature." There were times, however, when it seems he did manifest his attributes as the second person of the Godhead. For example, when confronting the Pharisees, Jesus replied while "knowing their thoughts" (Matthew 12:25).

Near-Death Experiences

A print in Flammarion's 1888 book *L'atmosphère* shows a man with his head stuck through the supposed curtain separating earth from heaven. His attempt to reach heaven via an alternate route worked! The title of the old country song gets

---

[279] Jesus's "self-emptying" of some aspects of deity is refered to as his *kenosis*.

it right: "Everybody Wants to Go to Heaven, But Nobody Wants to Die."

Recently, two best-selling books have related stories about two different boys' "death" experiences and subsequent returns to earth (both written by their fathers).[280] There are literally dozens of similar "heavenly tourism" books recounting near-death experiences (NDE), many from evangelical publishers, all describing someone's journey to heaven and back.

Those who study NDEs say there are two questions to ask about the experience: Are the subjects truthful and are their perception as they seem? I don't doubt that many experiencers are describing their real perceptions. But when paranormal explanations are offered, it's always prudent to examine the nature of the perception. Is it possible that NDEs have physiological or psychological explanations?

Neuroscientists suggest that the common "bright light" at the end of a tunnel is a sensation caused by decreased blood flow in the visual centers of the brain. (I was once advised by my surgeon: "If you see a bright light, run the other way!") Could natural processes also bring about detailed images of a place with people and angels? I'm not sure, but I know dreams aren't necessarily supernatural, and my dreams are quite detailed at times.

And to be scientifically precise, no dead person since the biblical accounts has truly returned to life. There is a qualitative difference between brain death (biological death) and states of various organ cessations (clinical death). "Near-death

[280] *The Boy Who Came Back from Heaven* (2010) recounts Alex Malarkey's NDE. In 2015 Alex released this statement: "I said I went to heaven because I thought it would get me attention. When I made the claims that I did, I had never read the Bible. People have profited from lies, and continue to. They should read the Bible, which is enough. The Bible is the only source of truth. Anything written by man cannot be infallible." (Jocelyn McClurg, USA Today, January 16, 2015.) The incident caused a firestorm of controversy among Christian publishers and distributers, some of which kept the book in circulation long after learning the truth. Obviously, this doesn't necessarily negate all such accounts, but it does show how vigilant we should be in discerning extrabiblical material.

experiences" are rightly named in that the victim has not actually reached biological death. Even medical professionals may not discern true biological death. When a victim flatlines and the attendants have a body that is so damaged resuscitation isn't expected, biological death is often declared with no EEG or further testing. When Jesus reanimated his friend Lazarus after four days, Lazarus was *dead* dead ("he stinketh," KJV). It's very unlikely that any NDEer has experienced biological death.

But let's assume that near-death is just that—*near* death. NDEers seem to suggest that the interface between this life and the next is a "fuzzy" place with choices of going forward or backward or lingering for a while. This doesn't mesh with Paul's picture of "absent from the body, but present with the Lord." However, most NDEers claim that there are corridors, bridges, highways, and all manner of labyrinths to heaven. (I'm thinking that when I die, I don't want an obstacle course. I would really just like to wake up in Christ's presence.)

NDEs were brought into mainstream culture in the 1970s with the book *Life After Life,* a summary of NDE anecdotes collected by M.D. Raymond Moody. One of the most influential books on the subject, however, was Mary Eadie's *Embraced by the Light*, released in 1992. During surgery, Eadie experienced being released from her physical body by three angels who transported her through corridors and eventually into the arms of Jesus. She claims to have sensed an outpouring of love *and* a high-speed data download. Eadie learned that Jesus and God were two completely separate beings and that her spirit had lived before physical birth, aiding God with the creation and the directions of her physical life. (Both of these doctrines are not mainline Christian tenets but are consistent with Eadie's Mormon faith.)

Among many other non-Christian beliefs, Eadie discovered that all religions were true and necessary because people have different spiritual needs. She was also told that her death was premature and that she needed to return to finish her mission (although to avoid complications, it would be necessary to erase what she had learned about that mission).

At the risk of overgeneralizing, it's interesting that in most, if not all NDEs, there is no evangelistic component. One would think that such a spectacular experience would compel the subject to urge others to do everything possible to prepare for the destination. On the contrary, the message is generally one of universalism—we all will end up in heaven regardless of our earthly decisions. The comprehensive teaching is "Chill out, God's our buddy and it's all okay!"

As with Eadie's LDS background, when NDEs from different cultures are considered we find that the belief system of the subject determines much about the experience. Buddhists move to nirvana. Pantheists become one with the cosmos. I suppose atheists find themselves dissolving into nothingness. There are some aspects, though, that seem to be universal. An article in *The Atlantic* summarizes some common elements of NDEs…

- The sensation of floating up and viewing the scene around one's unconscious body.
- Spending time in a beautiful, otherworldly realm.
- Meeting spiritual beings.
- A loving presence that some call God.
- Encountering long-lost relatives or friends.
- Recalling scenes from one's life.
- A sense of connectedness to all creation. [281]

Of course, there are stories of NDEs that are more terrifying. Some have claimed to arrive at the gates of hell rather than heaven. The accounts usually end with the subject turning to God in repentance. I'm not sure what to make of these, since

---

[281] Gideon Lichfield, "The Science of Near-Death Experiences," www.theatlantic.com, Accessed 1 Apr 2020.

I doubt Satan would desire that outcome. And if God was the source of the supernatural preview, it certainly diminishes the need for spreading the gospel here on earth. The rich man's brothers in the parable had "Moses and the prophets," implying that Scripture is enough.

I can understand a Christian's desire for NDEs to be true. After all, we know by God's revealed Word that Heaven *is* a real place. I wouldn't fault anyone for wanting to believe the validity of a message from someone who is overjoyed at having seen heaven. But this is certainly a place to ask about God: "Could he? Would he?" And as with all paranormal phenomena, we should compare the experiences with the revelations in Scripture, and not our desires and presuppositions. If the NDEer returns with new "facts" about heaven, we should be especially skeptical. As Hitchcock explains…

> …direct, divine, authoritative revelation ended with the conclusion of the canon of Scripture… Therefore, claims of "new truth" or revelations from God must come from another source, be it fallible human or deceptively demonic. [282]

I assume that demonically-influenced NDEs would be as feasible as other demonic phenomena.

Even when considering only the NDEs of those who claim a Christian worldview, we find a wide variety of descriptions. Many are mutually contradictory which means some (or all) can't be correct. But even if all the accounts were consistent, we must ask whether we need more information than what Scripture gives us. Only four people in all the Bible were ever allowed to see "beyond the curtain."[283] In each of their

---

[282] Mark Hitchcock and Jeff Kinley, *The Coming Apostasy: Exposing the Sabotage of Christianity from Within*, (Carol Stream, IL: Tyndale Momentum, 2017), 163.
[283] Isaiah and Ezekiel in the Old Testament, John and Paul in the New. Stephen and Micaiah were also allowed very brief glimpses.

accounts, we get very little information about the divine realm. Perhaps Paul gives the reason. What he saw were "inexpressible things, things that no one is permitted to tell."[284] Some NDEers also confess that the sights are beyond words then proceed to write many words about the experience. If Paul couldn't explain heaven, that speaks volumes. Hank Hanegraff sees the irony…

> It is simply incredible to think that Christians willingly accept the paltry substitutes offered up by modern near-death experiencers.[285]

Could God be blessing us today with an abundance of new extra-biblical details about our future home? Are we getting fresh insight to stir our imaginations? Spurgeon would think otherwise…

> …never think that imagination can picture heaven. When it is most sublime, when it is freest from the dust of earth, when it is carried up by the greatest knowledge, and kept steady by the most extreme caution, imagination cannot picture heaven. "It hath not entered the heart of man, the things which God hath prepared for them that love him." Imagination is good, but not to picture to us heaven. [286]

I would never fault a child or any other sincere person for having an interpretation of an experience. The subjects have often gone through traumatic and often agonizing recoveries. But I believe we are allowed to hold any perceptions up to the light of God's Word. As far as we can tell, the individuals from

---

[284] 2 Corinthians 12:4. Paul's appraisal: "…whether in the body or apart from the body I do not know, but God knows."
[285] Hank Hanegraaff, "Heaven Is Real," www.equip.org, Article ID. HIR20140418, 16 Apr 2014.
[286] Charles Hadden Spurgeon, *Spurgeon's Sermons Volume 02*: *1856*, Sermon 56, 22.

Scripture who were raised from the dead didn't arrive with new revelations about the spirit realm. Should the future bring more NDEs (and it likely will), believers must apply scriptural criteria. Paul warns about new revelations…

> … that you may learn by us not to go beyond what is written (1 Corinthians 4:6).

No true experience, regardless of the sincerity or "good works" of the subject, will negate "even the smallest letter or mark" of God's Word.

Artificial Intelligence

The completely physical concept of a computer "brain" has always tried to stretch its cyber-tentacles into the paranormal realm. We tend to flirt shamelessly with the notion that our Pinocchios can become real boys that know us and maybe even have feelings for us. I believe the supernatural context of artificial intelligence (AI) originates in our innate sense that our own souls are partly supernatural. Surely if we can make a machine that communicates with us and seems to understand, it must be verging on paranormal, right? Isn't there some way to push it beyond the natural?

In 2016, Hanson Robotics activated Sophia, a state-of-the-art "synthetic human." Sci-fi nerds across the world went ballistic. Finally, we have a robot woman who can converse, make facial expressions, joke, and do just about anything a long-awaited android should do. Sophia instantly became the sweetheart of the media. The Saudis even gave her full citizenship. So, is she the real deal?

The answer depends on the question. Can she act like a human? Of course. That's why she was built. Is she conscious like a human? Facebook's AI guru Yann LeCun isn't convinced…

> …many people are being deceived into thinking that this (mechanically sophisticated) animatronic puppet is intelligent. It's not. It has no feeling, no opinions, and zero understanding of what it says. It's not hurt. It's a puppet. [287]

Essentially Sophia is a very expensive, well-interfaced computer (or more precisely, a computer program that can operate some really fancy peripherals). Just like Siri, Alexa, and all their cousins, Sophia's program can calculate and run a lot of IF/THEN statements very quickly. But AIs like these do raise some worldview questions.

We should probably distinguish between *intelligence* and *consciousness*. On one level, brains and computers are similar in the way they process data. Neuroscientists believe the brain has an incredible number of pathways for handling data, just as a computer has circuits. If this is what is meant by "intelligence," then yes, Sophia has intelligence. She and Siri are actually "thinking" in a basic sense. And it's conceivable that computers could be (and probably already are) smarter than humans. Does Sophia *know*? Does she possess *artificial consciousness*?

The question now becomes "What is consciousness?" Science has revealed much about how the brain functions, but precious little about how human consciousness comes about. We discussed earlier that the biblical view of our human "self" is that we non-physical spirits residing in physical "tents" (2 Corinthians 5:1-7). This is known as *substance dualism*—part of us is natural and part is supernatural. The opposing view is *non-reductive physicalism*, or that the soul somehow emerges from the physical workings of the brain. A few Christians

---

[287] Noel Sharkey, "Mama Mia It's Sophia: A Show Robot Or Dangerous Platform To Mislead?" www.forbes.com, 17 Nov 2018.

embrace this belief, but it does seem to have many theological problems.

A naturalistic, evolutionary worldview, though, *requires* physicalism as an explanation for consciousness. There is no other viable choice for naturalists. It's easy to see how many scientists (and the culture at large) are in anxious expectation of a fully sentient AI since most accept naturalism. If human consciousness is physical, then it's simply a matter of time until neuroscience can synthesize it in a machine. On the contrary, the biblical worldview sees consciousness as "breathed into" humans. Our "self" is a nonmaterial entity connected to, yet not dependent on the physical body. Thus, true consciousness can never be attributed to a manmade physical machine.

Several paranormal concepts are related to the idea of human "modification" or "re-creation." These ideas fall into the category of *transhumanism*, the merging of humans with non-human technologies. We see this at a low level with embryo selection and artificial body parts, but the die-hard transhumanist is much more ambitious. If non-reductive physicalism is true, then the sci-fi enthusiasts' sky is the limit. Elon Musk and his crew will be able to live forever, moving their physical "souls" into robots. We can inject a college degree into our kid's head instead of going into debt. (But on the downside, our android servants will eventually turn on us and form their own army.)

I'll state it again: *Worldviews have consequences!* How we deal with AI and other transhuman concepts will depend on how we regard the supernatural. Unfortunately, many decisions are being made today under the assumption that physicalism is equated with total reality. There is no place for a transcendent, immortal, supernatural spirit. What a sad and empty legacy for our young people. On a personal note, I have no fear of an AI takeover in the future. Machines may be able to think more/faster/clearer, but they will never be able to consciously will/ponder/dream. And if perchance I'm wrong—we only need to wait for the next "blue screen."

20
A New Age of Old Ideas

"The force is what gives a Jedi his power. It's an energy field created by all living things. It surrounds us and penetrates us. It binds the Galaxy together."
Obi-Wan Kenobi[288]

"When the moon is in the Seventh House,
And Jupiter aligns with Mars.
Then peace will guide the planets,
And love will steer the stars.
This is the dawning of the Age of Aquarius."
The Musical *Hair*[289]

Over the past thirty years in the UK, nearly 5 million cows have been euthanized in one of the largest disease eradication programs in history. BSE (mad cow disease) is caused by an obscure neural pathogen called a *prion*. One of the smallest disease-causing agents, a prion is simply a misfolded body protein with the ability to pass its altered structure to normal proteins in the brain and nervous system. The twisting of affected proteins makes them useless in their prescribed

---

[288] Obi-Wan Kenobi, Lucas, George, dir. *Star Wars Episode IV: A New Hope*. Twentieth Century Fox, 1977.
[289] James Rado and Gerome Ragni , *Hair*, 1967.

function, thus, as the aberration spreads, brain function is lost. A body's defense systems are unable to detect prions since they are so similar to normal proteins. This similarity also makes it impossible at the present to treat prion diseases.

What an apt metaphor for New Age religion. How does Christianity counter an ideology that is in many ways so superficially similar that it's often undetectable? Even when it can be identified, the next occurrence will have a different "twist" and may not resemble the first. And even though the twisting may seem minor, its effect is deadly. The functioning of the affected portion in the context of biblical truth is compromised to heresy.

Some claim that there are only two religions ever practiced by humankind—one that exalts the true Creator God and the other that dethrones him and exalts another being. That may be an oversimplification, but it certainly gives a good perspective on New Age thought (Eastern mysticism) and the religions it informs.

Studying the New Age may be compared to studying the concept of "love." A broad topic such as this would be experienced differently by a great number of people. There are different manifestations of love, such as romantic or platonic. Some people experience love more deeply than others. "I love my wife." "I love chocolate." If I use the word "love" in a certain context, the hearer may be thinking in a completely different context. In the same way, New Age religions are so varied and widespread that a concise, black-and-white definition is often evasive.

Many in the modern Eastern mysticism movement downplay the term "New Age" since it has taken on a negative connotation in some discussions. I'm using the term generally as a category for a wide variety of ideologies. Some may claim that these are merely related to New Age by common interests or that New Age should be included as just one branch of another belief system. But let's not concern ourselves with details at this point. This chapter will examine the basic tenets of the New Age as they relate to the paranormal.

Decloaking the New Age

Defining Eastern mysticism is very much like the nailing-jello-to-a-wall idiom. As with occult groups, when a general characterization is put forward for New Age, one of the constituent groups will protest: "Wait, that's not what we believe!" And there likely will be some truth in it. Also, much of "evangelical" mysticism (aimed at recruiting Westerners) is couched in familiar Christian terminology. New Age will often sound very Christian.

So, to get our minds in mystic mode, let's start with a list of some common themes, beliefs, or threads that often run through New Age doctrine. Remember, this is only a loose set of very general ideas and buzz words (not necessarily bad or exclusive to New Age) that seem to characterize the groups…

- everything is God
- you are God.
- the power within
- realize that you are God.
- mind over matter
- truth is relative.
- your truth: no better than mine
- tolerance
- self
- balance good with evil
- yin yang, rainbow
- higher levels of consciousness
- harmonic convergence
- ESP

- I'm okay, you're okay
- karma
- holistic
- thoughts shape reality
- your reality may not be my reality
- reincarnation, out-of-body
- a New Age is coming!
- visualize it
- paranormal
- age of Aquarius
- spirit guides, ascended masters
- Zen
- Mother Earth
- the Force is with you
- pagan, primal
- create your own world.
- astral projection
- Christianity = limitation
- Western = bad
- yoga, chi, chakra
- Reiki, Feng Shui
- avatars, Nirvana, Krishna
- life energy, peace, love, happiness
- intuition
- human potential
- transcendental
- evolving
- channeling, séance, crossing over
- indigenous peoples have it right
- alternative
- name it, claim it
- enlightenment, higher plane
- auras, vibrations, crystals, oneness
- astrology, psychic reading
- Lucifer the Light

After reading this list, you may be somewhat put off over some of the items. Hold your fire! That's expected. Again, not everything noted is evil or exclusive to New Age. (Although I must say that if a significant number of these offend you, you may be deeper into New Age than you might think.) Inclusiveness, mimicry, and adaptability are characteristics of New Age spirituality. These are strategies aimed at an objective…

> What makes [an activity] part of the New Spirituality is not necessarily the practice itself… but the goal of the activity.[290]

The New Age belief system is far from new. Satan, the Father of Lies, initiated the first tenet when he put forth "I will be like the Most High." The implication is that one *can* be like the Most High, and there has been an unbroken chain of this rebellion up to the present. New Agers are named as such due to their anticipating and striving toward a "new age" in which humankind finally realizes its deity and rises to its well-deserved higher plane of existence. This is both a personal and a corporate quest: *I can become enlightened and we all must eventually become enlightened.*

All is One

After Satan's ancient quest for godhood, the earliest inklings of this same notion emerged in the pantheistic beliefs of Sikhism, Hinduism, Confucianism, Taoism, and some aspects of Buddhism. The basic tenet of pantheism is *monism* which claims that reality consists of only one essential

[290] Glenn Sunshine, "New Age Practices and the New Spirituality, Emerging Worldviews 19," www.breakpoint.org, 19 Nov 2019.

substance (thing). In other words, minds, matter, energy, God, the universe—everything—is essentially just *mind*. Christianity is *dualistic* which means that we believe separate things exist, specifically minds *and* a separate material world. With this simplified explanation, let's examine some aspects of New Age thought. I've spoken with and read many New Agers over the years. In the following section, I'll attempt to give the aggregate of ideas from the perspective of a typical proponent of New Age philosophy…

"Humans, snails, trees, bacteria—our souls are all joined spiritually with each other. Like cells in a body, all organisms are "cells" in a spiritual body. When part of the body suffers, we all suffer. And in a way, isn't Planet Earth like a big organism with many smaller living parts? It's like Mother Earth has her own spirit (we'll call her Gaia) so we should treat her with great respect."

"We are all divine, little gods. God is everything. Everything is God. If there is a supreme god, it's only our combined spirits acting in unison to create some impersonal, benevolent, magical "force" (yes, like Star Wars) that can be used if we can become enlightened (yes, like Jedi Masters). We can speed up the process by practicing New Age techniques of enlightenment, which would obviously start with shedding our old, traditional beliefs. That's what Jesus did, right? He showed us how to dump tradition and live a good life in order to find the divine within. He achieved the Christ-spirit. So can we!"

"Why should we listen to religious dogma or old stories? We are the Divine! Truth is what we think it is. No, it's what I think it is. Our truth may change as we evolve, but no problem—that will be our new truth. If we happen to see different or conflicting truths right now, that's okay. We're all at different places in our spiritual journey. We must be tolerant of others' ideas because all ideas are equal and good. And naturally, if we all are living by the truth we believe, we're all good people! Unless of course, your ideas are intolerant, and you refuse to

see our enlightened ideas. It's bad that you're so far down on the evolutionary ladder!"

"The universe is important because it's god—well, what we perceive as the universe. After all, we are one big mind. So, if I can visualize something, it will actually be there. Well, as 'actually' as anything else. Anyway, it doesn't truly matter what the truth is as long as someone believes it. In the meantime, we should take care of the world we see around us. Those whales and spruce trees will eventually evolve to meet us. Or have the whales already passed us? I'm not sure..."

Most thinking people can easily see the logical inconsistencies in New Age ideology. However, most New Agers place little or no priority on logic. That's not my biased criticism, it's the nature of their philosophy (or non-philosophy). Logic, rationality, and reasoning are only products of the mind considering it, thus, they have no true, underlying reality.

A logical thinker, though, must ask many questions... What's the ultimate purpose of the New Age universe? Why do we still have a strong moral sense of right and wrong if it's only an illusion? What happens when my perception of reality clashes with yours? If I create my own reality, why can't I do a better job of it? New Agers attempt to provide answers, but none are convincing. By contrast, the Christian worldview is based on absolute truth derived from the single reality that is God. It offers a logical, coherent explanation for the logical, coherent universe we observe.

New Age ideas will typically be intertwined with other worldviews. Its very essence is to merge and find unity with all. Therein lies much of the problem. Like a chameleon, New Age is often cloaked in the terminology of other thought systems. As previously stated, New Age concepts have crept into Christianity by the crafty exchange of terms so that they may speak of Christ, heaven, salvation, and still slip in meanings that are far from what the undiscerning would recognize.

Spirit Guides

Where Christianity centers on redemption through Christ, New Age is all about *enlightenment.* The central dogma is opening the mind's eyes to the fact that we are gods, allowing us to mature and evolve toward higher planes. But knowledge that enlightens must come through a channel. This conduit might be self-reflection (meditation), mind-altering drugs, psychic energies, or teachings.

Most branches of the New Age embrace the concept of *spirit guides*. These are spiritual personalities who may be contacted for guidance and wisdom. The origin of these entities varies among proponents, but most see them as *ascended masters*, those who have reached a higher plane of existence (usually through a culminating physical death after many reincarnations). A variety of spirit guides have been "adopted" by New Agers. The most popular by far is Jesus, but Buddha, Mohammed, Moses, Michael, and Lucifer have their own following.[291]

One popular New Age evangelist has been channeling her spirit guide for almost fifty years now. This entity claims to have lived in the lost land of Lemuria thousands of years ago. As a great military leader, he fought against Atlantis. If warriors aren't your thing, the chic trend today is to have an extraterrestrial spirit guide who can mind-meld from a distant galaxy. Newcomers to the New Age, however, may want to

---

[291] The "Jesus" named is obviously a caricature fitting the proponent's concept of Jesus. He (it) is often referred to as the *Christ Spirit*. The idea is that the "true" Jesus (the man) was so enlightened that he attained the highest level—realizing he was completely God. This subtle lie is so deceptive, that many Christians rework their biblical worldview to fit it. The idea that we too can become enlightened and take on the Christ Spirit echoes the deception of Eden's serpent.

start smaller. If you're lucky, you can channel one of your ascended master ancestors.

Frivolity aside, I hope the *extreme* danger of contacting spirit guides is obvious. What a demonic holiday it is when a person willfully and enthusiastically gives an open invitation to "anyone out there." I believe many modern necromancers (dabblers and pros) *are truly receiving supernatural messages*. Satan has his own doctrines and anxious "faculty members."[292] This is a biblical principle, thus the admonition to "test the spirits." The warning also applies to the "secondhand" hearers of those channeling the messages. Only God's mercy stands between the channeler and any lie the spirits may dish out.[293] Could this be another reason God forbids contacting the dead?

> There shall not be found among you anyone who... practices divination or tells fortunes or interprets omens, or a sorcerer or a charmer or a medium or a necromancer or one who inquires of the dead, for whoever does these things is an abomination to the LORD (Deuteronomy 18:10-12).

Kabbalah

Typical of other New Age belief systems, Kabbalah is difficult to pin down with definitions and doctrines. Its philosophy has influenced many thinkers such as Soloveitchik, Jung, and Freud. Surprisingly, the ancient, esoteric philosophy derived from Jewish mysticism is especially popular among celebs such as Madonna, Roseanne, and Streisand (beware if you're a one-name pop icon!). Many of the writings and tenets

---

[292] "...some will depart from the faith by devoting themselves to deceitful spirits and *doctrines of demons*" (1 Timothy 4:1).

[293] This is the antithesis of Christian meditation. Yes, we listen for the Spirit's leading, but only in association with immersion in God's written Word.

are very subjective and attempting to follow them logically is difficult. Here are a few ideas that characterize the general themes of Kabbalah…

1. God is a source of infinite spiritual light.
2. Human souls existed before coming to earth.
3. God poured some of his light into humans.
4. Some human "vessels" broke. Their light scattered and became imperfect.
5. The Kabbalah was given to the angels, Adam, Abraham, and finally Moses, in order to teach us the pathway back to perfection.
6. Our job as humans is to help God to restore the Universe to perfection.
7. The sin of Adam and Eve was the failure to bring the universe to a new level of consciousness.
8. Four primary archangels deal with humans: Michael, Gabriel, Uriel, and Raphael.
9. There are two distinct Messiahs, one from Joseph's line and the other from David's.
10. We can, and should, develop a direct experience with God by learning the principles of Kabbalah.

Kabbalah teaches that the ultimate god (*Ein Sof*) is unknowable, but that he reveals himself in 10 forms, or *sefirot*. Each sefirot represents one component of the total entity. They are often shown in a *tree of life* diagram with connections between certain sefirot. A lesson we can learn from Kabbalah is that outdated, and mostly incomprehensible ideologies can still be attractive if dressed in New Age promises of power and enlightenment.

Reincarnation and Karma

Several early Greek philosophers entertained the idea that a soul could move from one physical body to another. *Spirit transmigration* has primarily been a belief within Hinduism, Buddhism, and other Eastern religions. The belief sneaked into other cultures in the 19th century through the teachings of influential Transcendentalist and Theosophy leaders. Reincarnation also makes a good TV/movie plot, as Hollywood has given us *A Dog's Purpose*, *My Mother the Car*, and several *Three Stooges* episodes with transmigration themes. Even General George Patton believed he was the reincarnation of the military hero Hannibal. There are also a few within Judaism and liberal Christianity who have courted the belief, but my concern in this section is with the general "westernized" take on reincarnation that has been gaining popularity and seeping into some Bible-believing churches.

Reincarnation is closely linked to the concept of *karma*. There are many variations on this belief, but generally, karma involves the accumulation of good or bad deeds in a lifetime and how these are resolved.[294] Karma is similar to Scriptural truths in that it implies a need for some type of payback or justice. Perhaps that's part of our *imago dei*. We intrinsically know right from wrong and that wrongs should be righted. The principle of karma, however, says that if the wrongs aren't sufficiently balanced in one life, there must be subsequent lives in order to obtain justice.

In theory, a person's spirit (the true self) will be reincarnated into physical bodies as many times as it takes for bad deeds to be balanced. Some proponents believe the new body could be any living thing—human, plant, or animal. Hopefully, if or when karma is completely satisfied, the soul

---

[294] Interesting, since *good* and *bad* are relative or erroneous concepts to most New Agers.

will now be "free from [re]birth" (as George Harrison sang) and it can now rest in the nothingness of "godhood."

Ultimately, reincarnation is a nondual philosophy that sees only the spiritual to be real or important, and the fleshly as a deception. Some claim this coincides with the biblical view of spirit and flesh. An accurate assessment, however, reveals that even though the spiritual is of highest priority, God also places value on his physical creation. There are some logical problems with the actual workings of reincarnation and karma, but to a proponent, those discrepancies are physical, and the physical is unimportant.

In the final analysis, reincarnation and karma stand in opposition to the Christian worldview. Scripture tells of a single, unique *in*carnation. We believe that the Second Person of the Trinity did, at a point in time, take on a fleshly, human body and that that body was raised and will be the physical manifestation of Christ through eternity. We also believe that our spirits will be joined to our perfected bodies in a process of *resurrection*, not bounced from one organism to another in an erratic attempt to atone for our bad deeds. Our dual spirit-bodies, designs of our Creator, will live in communion with our Creator with our sins atoned by Christ.

> And just as it is appointed for man to die once, and after that comes judgment, so Christ, having been offered once to bear the sins of many, will appear a second time, not to deal with sin but to save those who are eagerly waiting for him (Hebrew 9:27-28).

Yoga

I served several years as chief lawn mower for my family. My arch-nemeses during that time were bumblebees in a hive at the far edge of the yard. Unlike honeybees, bumbles have underground colonies with a single small opening at the surface. From experience, I learned they are mean, painful, and dislike having their entrance mowed over. Bumbles are relentless. They will chase you down regardless of how far you run. However, I discovered that as long as the mower didn't get too close, they were okay with mowing.

That would have been the end of the problem had I not been obsessive/compulsive about mowing. The unmowed clump on the edge unnerved me. I could have left it and no one would have ever noticed but that wasn't the point. I wouldn't be bested by insects. Every time I mowed, I trimmed as close to the hole as possible. Soon I had the clump down to a minimum with only a few angry bumbles.

Many Christians have a similar relationship with the New Age. Some of its elements are undisputedly pagan-supernatural yet are essentially harmless at a certain distance, at least in a personal sense. As long as I maintain a safe distance, all is well. The question then becomes "How close is too close?" Is the gain worth the risk of stings? Regarding the "grayish" aspects of New Age, I'm often asked whether this is a hill we need to die on. We certainly don't wish to burn evangelistic bridges, but we must ask in response, at what point is the body count too high?

A good model for this discussion is yoga. Many, including Christians, practice yoga regularly with no perceivable, immediate harm, in fact, most seem to benefit. So, what's the problem? No one denies that yoga has its roots firmly planted in Hindu/Buddhist pantheism. The simplified objective of all flavors of pantheism is for an individual to realize his or her "oneness" with the rest of the universe. Yoga as originally designed was not intended to be physical exercise, but a way of posing the body for the efficient flow of "life energy."[295] An offshoot of this is the belief that one's creative life force, the

[295] We see the same *life energy* concept in Feng Shui and some alternative health practices.

Kundalini spirit, lies coiled at the base of the spine. (Kundalini translates “serpent.”) True yoga teaches that by achieving the proper poses, the kundalini can awaken and release its energy…

> When this sleeping Kundalini is awakened it raises its [serpentine] hood. The door of the Sushumna [a *nadi* or energy conduit] is opened and the Kundalini ascends upwards along the Sushumna piercing through the six chakras… situated in it. When it reaches the highest centre, called Sahasrara, in the crown of the head, it unites with the Lord Shiva. [296]

Needless to say, Christians are divided on the nature of yoga. Can the practice be yanked from the occult, shaken free of its original intent, and sanctified for God’s use? Former yoga instructor and Reiki master Jessica Smith says no…

> I find no example in scripture where the Lord suggests that a good way to get closer to him or to honor him is to find a popular spiritual pagan practice that was created and developed for the purpose of worshiping and seeking communion with false gods, and just plugging his name in.[297]

Others believe the physical part of yoga can be separated from the spiritual…

> There is nothing wrong with stretching… The problem comes when one incorporates Eastern "meditation" techniques or other inherently religious practices that distinguish yoga from mere recreational stretching. [298]

---

[296] Elliot Miller, “Approaches to Yoga: Kundalini Yoga,” www.equip.org, 6 Apr 2009.
[297] Jessica Smith, www.truthbehindyoga.com, Accessed 28 Mar 2020.
[298] Jeremy Butler, “Should Christians practice Yoga?” www.carm.org, 15 Nov 2018.

Yoga proponents and instructors, however, teach that the philosophy of yoga is bound with the practice of asanas (poses)…

> What makes [gymnastics and aerobics] different from a true yogic practice is that their focus is only on the body. What makes a true yogic practice unique is that its focus is on a sustained feeling of freedom and wholeness.[299]

Again, the broader question here may be, how close is too close? Yoga gurus agree that repeatedly "going through the motions" (literally) leads deeper into the spiritual spiral. The idea of neutrality and the separation of components is foreign to the philosophy of pantheistic yoga. We are also prone as humans to be lured into "baby steps." A majority of Westerners who now practice true spiritual yoga began their journey as merely an exercise regimen…

> A trapper doesn't put a pile of dirt in a snare. He puts something tempting and appealing. [300]

This principle of spiritual desensitization is often seen in yoga as well as other New Age practices.

Should one wish to benefit from the physical discipline of yoga, are there alternatives? Smith says there are many alternatives to yoga that can provide the same physical benefits. If one has prayerfully considered the pitfalls of mysticism and still believes a safe distance is possible, those with experience in the area recommend the following guidelines for Christians who wish to continue practicing physical yoga…

---

[299] Alanna Kaivalya, "How We Got Here: Where Yoga Poses Come From," www.huffpost.com, Updated 28 June 2012.
[300] Jessica Smith, www.truthbehindyoga.com, Accessed 28 Mar 2020.

- Avoid the philosophy and vocabulary associated with yoga.
- Pray for wisdom and a clear mind while practicing yoga.
- Use relationships with others in your classes or sessions to thoughtfully spread truth rather than supporting error.
- Take care that "weaker brethren" are not being negatively influenced by your freedom to practice yoga (Romans 14). Could someone less disciplined be led by your actions, then lured into pagan philosophy?

Meditation

No sooner does the Book of Psalms get started than it encourages us: "Blessed is the one… whose delight is in the law of the LORD, and who meditates on his law day and night."[301] We are also told: "Keep this Book of the Law always on your lips; meditate on it day and night, so that you may be careful to do everything written in it."[302]

Unlike yoga, meditation has a scriptural precedent, yet many Christians are leery of meditation due to its "hijacking" by mystics. A comparison of New Age and biblical ideologies shows quite distinctively the importance of worldviews. We've seen that New Age groups frequently share some worldview elements and vocabulary with Christianity. Much of this is by design in an ongoing attempt to make mysticism palatable to those from Western cultures. *Christ* to a New Ager means the *christ spirit*, not *Jesus the Christ*. Christians may hear a mystic

---

[301] Psalm 1:1-2
[302] Joshua 1:8

say *Holy Spirit*, yet their meaning is the impersonal, universal *Brahma*.

Meditation has a similar issue with meaning. Both mystic meditation and New Age *mindfulness* techniques draw on the pantheistic concept of inner searching. These practices have a singular purpose…

> Eastern meditation… employs techniques like concentrating on objects, exercising "controlled" breathing, and uttering mantras, all this with the purpose of emptying the mind of all kinds of distractions which supposedly will enable a person to reach an altered state of consciousness... [303]

This type of mental focusing is ubiquitous in New Age ideology. The spiritual goal of the New Ager is to continually rise to "higher levels of consciousness" which requires less rationalizing and more feeling and experiencing. Propositional knowledge (the real world) must be cleared away so that esoteric, "cosmic" knowledge may flourish.

Biblical meditation is just the opposite—we are encouraged to *fill* our minds. Whereas New Age meditation and mindfulness look *inward* to find oneness, wisdom, and connection, scriptural meditation looks outward and *toward* an absolute truth, specifically the *Person* who is truth. The Psalms are in a sense records of meditation. They give several truths for us to meditate such as the person of God, his works, and his promises. Scripture posits an empty mind as a dangerous place, open to ungodly ideas and possibly the whisperings of malicious spirits. It's unfortunate that in overreaction to Eastern mystic practices, many Christians shun true biblical meditation.

[303] Christian Research Institute, "Christian Meditation: Is Meditation Biblical?" www.equip.org, 17 Mar 2009.

Jesus recognized the benefit of getting away to a solitary place to pray and meditate. He made it a priority…

> … great crowds gathered to hear him and to be healed of their infirmities. But he would withdraw to desolate places and pray (Luke 5:15-16).

The Biblical word for *meditate* carries the idea of "chewing something slowly for a long time," much as a ruminant would graze early in the morning then re-chew the rest of the day. In our rushed society, it's often difficult to find a time and place for meditation, yet as Jesus shows it has intrinsic importance in the life of a believer. Paul urges us…

> Whatever is true, whatever is honorable, whatever is just, whatever is pure, whatever is lovely, whatever is commendable, if there is any excellence, if there is anything worthy of praise, think about these things (Philippians 4:8).

Biblical meditation could focus on a single attribute of God, but more often it's a deliberate act of searching and pondering a truth. Though the outcome is often peacefulness, the duration of meditation may be filled with crying out, complaining, and pleading. Far from an *emptying* of the mind Psalm 77 displays a diversity of emotional questioning, pondering, and reasoning…

> When I remember God, I moan;
> when I meditate, my spirit faints.
> You hold my eyelids open;
> I am so troubled that I cannot speak.
> I consider the days of old,
> the years long ago.
> I said, "Let me remember my song in the night;
> let me meditate in my heart."

Then my spirit made a diligent search:
"Will the Lord spurn forever,
and never again be favorable?
Has his steadfast love forever ceased?
Are his promises at an end for all time?
Has God forgotten to be gracious?
Has he in anger shut up his compassion?"
Then I said, "I will appeal to this,
to the years of the right hand of the Most High."
I will remember the deeds of the LORD;
yes, I will remember your wonders of old.
I will ponder all your work,
and meditate on your mighty deeds.

The New Age and UFO Religions

The more esoteric UFO/ET phenomena play out well in many New Age circles. As we saw in our discussion of alleged abductions, victims are often indoctrinated with standard New Age ideology. One of the most disturbing UFO religions was observed in the Heaven's Gate group led by Marshall Applewhite. Applewhite maintained his commune at a mansion in Rancho Santa Fe, California, and used video messages to spread his ideas.

Essentially New Age teachings with a sprinkling of Christian eschatology, Applewhite instructed his followers that they would soon ascend into the heavens where an extraterrestrial spaceship was waiting (parked out of sight behind the Hale-Bopp comet). The aliens would then transform their bodies and take them away to a better planet. Impatient with the lack of predicted activities, Applewhite revised his doctrine, saying the followers would need to shed their physical bodies first. In 1997, thirty-nine of the group committed suicide, hoping for the promised close encounter.

Raileanism, possibly the largest UFO religion with about 100,000 members, is very much alive worldwide. Founded in 1974 by Claude Vorilhon (now known as Raël), Raelians believe that an extraterrestrial species known as the *Elohim* created all life on Earth through advanced biological techniques 22,000 years ago.

The theology of Raelianism reads like a sci-fi version of the Bible. Much like the ideas of von Däniken and Stitchin, every event in Scripture is reinterpreted as a naturalistic, ET intervention. Thus, the philosophy is a type of New Age atheism in which the only hope of transcendence and afterlife is through naturalistic means. Those means include cloning, genetic manipulation, AI, and a conglomeration of other transhuman concepts. Raelians expect the Elohim to return to Earth soon, and hopefully fix the mess we humans have created.

Raelians have received much publicity by their political and social activism. In an odd mix of liberalism and New Age ideas, they aggressively support, genetic manipulation, gender and sexual freedom, nudity, animal rights, anti-Catholicism, tolerance, and hedonism.

Hundreds of other UFO religions exist around the world and are collectively one of the fastest-growing belief systems. This shouldn't be surprising since the best deception always takes into account the current state of the deceived. What better way is there to upgrade Satan's age-old strategy of dethroning the Most High? I believe we should expect a worldwide expansion of New Age beliefs coupled with transhumanism and ufology. Only a clear understanding of God's Word can produce a worldview capable of discerning these cunning deceptions.

21
Mega-Miracles

"Every teacher, every preacher with the very best intent
Found new ways to hide the mystery
replaced by common sense.
And to know you was to keep you in my pocket
So easy to hold." Nichole Nordeman[304]

"Creation is not a brute fact without meaning. It derives its meaning from the divine character and will. As the theater of God's redemptive activity, creation is not static, but is moving toward that goal established by decree before the foundation of the universe." Albert Mohler[305]

In our pondering of supernatural events, it's easy to let all the trees obscure the forest. C.S. Lewis speaks of the *Grand Miracle*. That the unbounded Creator of all that is, could for a period of time take on a limited physical body, dwell among us, then die and live again—this is a phenomenal supernatural intervention. The creation event itself could also be considered a grand miracle. A self-existent, self-sufficient Being with no deficiency or need of any kind brought into existence the

---

[304] Nichole Nordeman, *Who You Are*, Sparrow Records, 1998.
[305] Albert Mohler, (2010). "The Christian Worldview as Master Narrative: Creation, "www.albertmohler.com, 2010, Accessed 8 May 2020.

"otherness" of lesser beings and a universe for their temporary home.

Regrettably, many carry the name "Christian" who either deny or reduce to metaphor the Creation, the Incarnation, and the Resurrection. I also know there is a remnant who still fully believes in the supernatural reality of these three miracles.

The issue, however, is with the large number of believers on the fence of belief who wince at the mention of these *mega-miracles*. I use this term only from a human perspective since God is able to perform the largest miracle as easily as the smallest. And that's the point… Why do we pick and choose among miracles? If we believe God can work *any* miracle, isn't he able to work *any other* miracle?

As discussed earlier, this hypocrisy is at the heart of our current worldview problems. Many Christians *profess* belief in selected supernatural events simply because there is no satisfying alternative…

> The truth is that Christians affirm the incarnation because they have to—it defines Christianity.[306]

I would add, the same is true of the Resurrection. If we discard the true supernatural transformation of Jesus from dead to alive, we are absolutely without hope…

> And if Christ has not been raised, your faith is futile; you are still in your sins. Then those also who have fallen asleep in Christ are lost. If only for this life we have hope in Christ, we are of all people most to be pitied (1 Corinthians 15:17-19).

As a scientist, I encounter a variety of beliefs among my colleagues regarding origins. Thankfully, my co-workers all base their creation views on a solid, factual Christian

---

[306] Heiser, *Unseen*, 186.

worldview. I have crossed paths with many Christian scientists, however, who believe they are boxed into a compromised view because the supernatural elements of creation are simply *too* supernatural.

In some way, "God plus naturalistic laws" is more palatable as an ultimate source of reality than God alone. This view sees the supernatural as an "add-on" to the natural system. In essence, it says there is something that exists *beyond* or *outside of* God. This *something* somehow helps span the cognitive gap between natural and supernatural. Yet the problem remains—isn't this simply a more complicated type of supernatural? Not only is it contrary to Scriptural truths about God, it once again implies that some miracles are too difficult for God.

The Incarnation and Resurrection are beyond the scope of this book and my mind. (I would guess that anyone denying these pillars of Christianity, wouldn't likely be reading this book in the first place.) Let's focus, instead, on the creation and three other supernatural events that seem to be targets for skeptics and many Christians who fear their ridicule.

On the surface, these divine interventions may seem like side-show attractions, but I believe they are central in the Christian worldview. Coincidentally(?), they are all found in the first eleven chapters of Genesis. Perhaps the most controversial passages in the Bible (except for Christ's miracles), the accounts seem to be magnets for Bible critics and liberals who reject the supernatural. These eleven chapters are also the most disturbing to Christians whose worldview is lacking an adequate accommodation of the spiritual realm as revealed in Scripture. Are the events described in these passages too supernatural to be genuine? Can the 21st-century Christian worldview embrace even a part of these mega-miracles?

Creation

As a Christian and a science teacher, I've dealt with this topic extensively. The evidence ranges from the simple observation of the orderliness of nature to discerning the complex origin of information in a molecule of DNA. Chapter 1 of Romans reminds us that there is so much obvious evidence for a Creator in nature (*things that have been made*), that all who refuse to believe are "without excuse…"

> For his invisible attributes, namely, his eternal power and divine nature, have been clearly perceived, ever since the creation of the world, in the things that have been made. So they are without excuse. For although they knew God, they did not honor him as God or give thanks to him, but they became futile in their thinking, and their foolish hearts were darkened (Romans 1:20-21).

Note that the issue of nonbelief isn't due to a lack of evidence or even the misinterpretation of the evidence. The sad indictment is pronounced on those who *refuse to acknowledge* the Creator despite *knowing the evidence*. And the judgment has consequences in the here-and-now: *Their thinking became futile and their foolish hearts were darkened.*

The creation is too often portrayed as a death match between science and faith. In the 17th century, astronomer Johannes Kepler was one of many past scientists who had no issue with a universe of miraculous origin. Despite popular opinion, many contemporary scientists also see no discrepancy between science and Scripture.

And realistically, the *who* of creation is the easy part for Christians. It's the *how* that raises questions. Four primary views are common among scientists and others who consider origins. There are numerous subcategories, nuances, and overlaps, but we'll keep it simple…

1. Young Earth Creationism accepts the literal interpretation of events in Genesis. The days of creation were six, twenty-four-hour days. Earth's age is roughly 10,000-*ish* years. There was also a worldwide flood that brought about many features we observe on the earth's surface.

2. Old Earth Creationism asserts that the days in Genesis are figurative references to long periods of time (billions of years). God created slowly, perhaps intermittently, all that we observe today, on earth and in the cosmos. Living things, including humans, were supernaturally created at appropriate times and experienced death for millions of years before the fall of Adam.

3. Evolutionary Creationism (theistic evolution) is basically the same as old earth creationism except that proponents acknowledge God used the process of slow evolution to bring life forms from ancient to modern. Little or no supernatural intervention was needed except perhaps in the very beginning.

4. Naturalism (Darwinian evolution) assumes only the physical universe. No intelligent, personal creator was needed, only random evolutionary processes. This is presumed to take long time spans to bring us to the present.

I appreciate the hard work of so many Bible-believing scientists who sift through piles of data from biology, geology, cosmology, and other fields related to origins. Like everyone else, I have my own view of creation events, but I've heard compelling arguments for the beliefs listed above (and several others).[307] There are still many questions to answer regarding

---

[307] My personal view of origins would be considered *young earth*. I realize there are data that seem to indicate a universe of extremely old age. I also see evidence that seems to contradict this, thus I've yet to find a good scientific reason for concluding one or the other. Because of this, my belief derives more from theology than from

how God created, and I would hope a healthy dialog will eventually resolve some of these.

It's one thing, however, to base views on legitimate data, yet quite another to disregard or eliminate particular alternatives simply because of their ties to supernatural intervention. I have a great deal of respect for scientists who might disagree with my take on origins because they have data that seem to support their conclusions. My problem is with Christian scientists who profess the supernatural as one of the bases of Christian doctrine, then reject it arbitrarily in practical applications to real events in the physical realm.

For example, the most flagrantly supernatural creation scenario is a literal six-day event, about 6000 years ago, in which God brings from nothingness the entire functioning universe, with strata and stars appearing to be of great age. Whether this is exactly true or not, the point is that often opinions rest on the degree of supernatural intervention with which one is comfortable: "If we are uncomfortable with the supernatural appearance of age, let's discard that and add a few million years. If the miraculous arrival of millions of fully formed species is a problem, let's throw in slow evolution." Again, *God can do a big miracle as easily as a little one.* Beliefs should not be based on the *least supernatural* explanation.

The Fall

The event often referred to as the *fall of man* is little more than a fairy tale today, even in many churches. Any mention of a real Adam and Eve is met with ridicule if acknowledged at

---

science. In my opinion, young earth creationism is the only alternative that addresses the sin/death issue. According to Scripture (New and Old Testaments), death entered the universe by Adam's sin. There have been many attempts to circumvent this allegorically or etymologically, but the results are less than satisfying. And, I have absolutely no problem with God's supernatural intervention in any way he desires. I suppose I'll remain in the young earth camp until I find a good theological explanation for death without sin or a Savior with no need to save us.

all. Less than a third of American adults believe the couple even existed (the lowest in three decades).[308] John Schneider, a former professor of theology at Calvin College summarizes the view of many of the Christian intelligentsia…

> Evolution makes it pretty clear that in nature, and in the moral experience of human beings, there never was any such paradise to be lost… So Christians, I think, have a challenge, have a job on their hands to reformulate some of their tradition about human beginnings.[309]

But Scripture is unequivocal on the subject, and for good reason. Adam's sin was the precipitating cause for our present predicament (regardless of public opinion on how or whether it's possible). Most Christian doctrines are based on the historical veracity of this event. The need for Christ's atoning work is centered on the sin of an actual historical man…

> Therefore, just as sin came into the world through one man, and death through sin, and so death spread to all men because all sinned… much more will those who receive the abundance of grace and the free gift of righteousness reign in life through the one man Jesus Christ (Romans 5:12,17).

And in a real sense, the entire gospel message is negated by the absence of a historical Eden. As Albert Mohler writes…

> The denial of an historical Adam and Eve as the first parents of all humanity and the solitary first human

---

[308] Art Swift, "In U.S., Belief in Creationist View of Humans at New Low," www.gallup.com, May 22, 2017.
[309] Barbara Bradley Hagerty, "Evangelicals Question the Existence of Adam and Eve," NPR Morning Edition, August 9, 2011.

> pair severs the link between Adam and Christ which is so crucial to the Gospel.[310]

Jesus himself referenced the real-life Adam and Eve in explaining the significance of marriage vows (Matthew 19), and their son Abel in speaking of the death of prophets (Matthew 23). Like any good Jewish boy, Jesus was schooled in the Old Testament (Torah). Never in his teachings was there even a hint that certain parts were allegory or myth.

So, what's the deal with eating the forbidden fruit? Many good sources are available that detail the significance of the sin of Adam and Eve. Here are some doctrinal axioms derived from the event:

- Adam and Eve originally had an untainted relationship with God, who communed with them (likely as the visible second person of the Trinity) in Eden.
- The couple was in a state of innocence at this time.
- God, who has always desired a free, loving relationship with humans, gave only one restriction—not to eat from a particular tree.
- In a free, self-willed choice, Adam and Eve decided to disobey God.
- That choice (sin) broke their relationship with the Creator, who let them pursue their own will.
- The result was a deteriorated, sinful nature within, and a fouled, distorted universe in which they had to live.
- All that could come from a tainted population was more tainted individuals.

Many have attempted to allegorize this account or vilify God, but any feelings of injustice come from our corrupted understanding of God and the true repugnance of sin. My

---

[310] Dr. R. Albert Mohler Jr., "False Start? The Controversy Over Adam and Eve Heats Up," www.albertmohler.com, Accessed Feb 26, 2020.

inherited sin nature has trouble accepting the idea that I might have an inherited sin nature. The fall has implications that rub our ego and self-esteem the wrong way. Modern education, self-help, and psychology begin with the premise that we are "good down deep" not "sinful by nature." But Scripture is unbudging at this point. "For I know that good itself does not dwell in me" (Romans 7:18).

From a scientist's perspective, the fall has enormous implications and great explanatory power. I observe a world in which decay is the norm. Organisms die. Animals eat other animals. Pain dominates. And this isn't just a planetary phenomenon. As far as we can determine, the entire universe is slowly slipping into dark, cold entropy. Does this seem like a condition for which God would give the appraisal "It is good?"[311] Couldn't an omnipotent and omniscient Creator do better? Of course, he could—and he did. A metaphorical Eden at best leaves us struggling for a theological reason for why things are the way they are. At worst it strips the Almighty of the very attributes that make him almighty.

Skeptics are correct on one account. A historical paradise and subsequent fall require a large dose of supernatural. If we accept that the universe is now in a fallen state, we must also postulate an original state. This is problematic for naturalists and Christians who balk at divine intervention.

It *is* difficult to imagine an *un*fallen condition. My fallen brain has experienced only a fallen world with rotting things, sick things, and messed up things. Could there be a system in which death isn't a factor? Can an ecosystem exist without a predator? Can energy equations find an endless input of new energy? *The answers are yes, but they will forever remain faith-based in this physical world.*

Again, my goal isn't to prove this, but to lay it out as a reality that Scripture presents as happening both in our past and

---

[311] Genesis 1.

in our future. No, I don't understand how Eden's food chains functioned. I don't know whether erosion occurred. I can't explain what major changes must have occurred so that a non-predator becomes a predator. I can only imagine it was a phenomenal and horrendous event. But isn't that exactly what Scripture portrays? One of the longest threads running through the Bible is that sin is terrible, and its effects are monumental—so monumental that the Creator himself had to die. My Christian worldview demands an explanation of how things were, how they got the way they are, and how they will one day be.

I'm thankful that God in mercy didn't end the Story here. Before the curtain falls on the tragic scene of humankind's fall, the Creator promises a rescue. One is foreshadowed who will come and "bruise the serpent's head." It's comforting to know that the effects of the fall will be supernaturally reversed someday. Without understanding completely, I anxiously look forward to the day when…

> The wolf shall dwell with the lamb, and the leopard shall lie down with the young goat, and the calf and the lion and the fattened calf together, and a little child shall lead them. The cow and the bear shall graze, their young shall lie down together, and the lion shall eat straw like the ox. (Isaiah 11:6-7).

The Flood

The story of Noah and the flood is so etched into our minds from storybooks and coloring pages that our natural inclination is to think myth or fairy tale. After all, two-by-two, floating zoo, rainbows—it seems better suited for a bedtime story than for history.

But the flood story is one of the ugliest in Scripture—not just the eventual destruction of multitudes, but the debauchery that led the Creator to "regret" that he had created. Clearly, the flood account overall is a foreshadowing of Christ's work. We are told as much in 1 Peter 3. God's immediate purpose, however, was straightforward…

> I am going to bring floodwaters on the earth to destroy all life under the heavens, every creature that has the breath of life in it. Everything on earth will perish (Genesis 6:17).

The incidents leading God to make this judgment are discussed in another chapter, thus, we'll examine only the consequences here.

Skeptics dismiss the flood account for two primary reasons. First, God's actions seem extreme. Wasn't it "overkill" for a merciful God to wipe the slate clean—literally? Again, our fallen minds have difficulty discerning the horrid nature of sin, but we'll deal with this aspect later, as well.

Second, as with the other Genesis events, the account seems *too* supernatural. Every attempt has been made to minimize or naturalize the flood account, but the results tend to make the incidents more irrational since they also diminish the original purpose. Rather than analyze all of the evidence for a worldwide flood, let's focus on the supernatural aspects that come into question.[312]

Was the flood worldwide? Some Bible scholars point to the fact that if God's intent was to destroy sinful people, perhaps he could have flooded only the Mesopotamian region since, arguably, that's where the human population of the time lived. I see the logic of that idea, but it's contrary to what a literal interpretation of Scripture provides. Also, if the flood was a

---

[312] For full analyses of the young-earth perspective of flood events, these sites provide extensive resources: www.creation.com, www.answersingenesis.org, www.icr.org

local event, why require the extensive ark project? Why not just have Noah's clan and the animals take a trip upland?

Is there enough water to cover the earth? Or as skeptics smugly frame the issue: There isn't enough water on planet earth to cover Mt. Everest. This brings up an important concept that is often pushed aside: *The flood was a monumental cataclysm...* Our tendency is to think relative to present-day floods, but that would be contrary to all that Scripture tells us about the deluge. *God can cause a worldwide flood as easily as a local flood.* Genesis 7:11-12 describes the actions of the flood...

> ...the same day were all the fountains of the great deep broken up, and the windows of heaven were opened. And the rain was upon the earth forty days and forty nights (KJV).

Morris explains...

> The reference to "broken up" merits attention, for it implies a wrenching of the earth's crust, a great tectonic event. The same word is used in Numbers 16:30-33 to describe the supernatural opening up of a great pit into which the rebellious Korah and his followers and their families fell...[313]

The flood was a *catastrophic* event sent to judge a *catastrophic* condition. We don't know the exact changes the flood wrought on the earth's surface, but there is no reason to set limitations on it based on the features we presently observe. Even after the deluge, God continued to supernaturally reform the landscape...

[313] John D. Morris, "All the Fountains of the Deep," www.icr.org, 11 Aug 2012.

> You covered [Earth] with the deep as with a garment; the waters stood above the mountains. At your rebuke they fled; at the sound of your thunder they took to flight. The mountains rose, the valleys sank down to the place that you appointed for them. You set a boundary that they may not pass, so that they might not again cover the earth (Psalm 104:6-9).

Was the ark project feasible without supernatural intervention? The question divulges the irony. Why do we continue to push as far away from the supernatural as possible? I applaud those who do research on the physical parameters of the ark and provide information about what a colossal undertaking the project must have been. Ark replicas and museums are wonderful ways to visualize the vessel and imagine the awesomeness of the scene. Good apologetics can allow us to marvel at the account, but we don't need to "prove" the ark story naturalistically in order to "prove" Scripture.

The veracity of the Scripture stands on its own legs. The lessons it teaches about the flood are about God himself. I believe we should apply the same standards to mega-miracles, such as the flood account, as we do to any biblical miracle. Will God always use the most miraculous methods possible? Usually not. By faith, we accept the supernatural component as validation of the message.

The flood ordeal exhibits a mixture of human and divine action. God could have miraculously protected eight people from the waters as he protected three from the fire of the furnace, yet he involved Noah's family for decades in constructing the ark. As far as we can tell, the family fed and cared for the animals, but it was a supernatural call that brought them to the ark. God *is* supernatural and *can* intervene supernaturally as he pleases.

I believe the best way to view the flood account is simply to take it as Scripture offers, applying natural and supernatural where and how they are presented. This is the precedent that Jesus himself set when referring to Noah and the flood. Any attempt to do otherwise would not only contradict the words of Jesus but would also negate the spiritual truths he was illustrating.

Babel

When the topic of the Tower of Babel comes up, skeptics rant about the foolishness of believing that God supernaturally "manufactured" all of the languages on earth at one time. Yet the linguistic aspect is eclipsed by the ensuing spiritual drama. The account begins in Genesis 11 with a simple summary of the period of time since the flood...[314]

> Now the whole earth had one language and the same words. And as people migrated from the east, they found a plain in the land of Shinar and settled there (Genesis 11:1-2).

Geographically, Shinar (Mesopotamia) was roughly modern-day Iraq, Kuwait, Syria, and part of Turkey. It was a major portion of what has been called the "fertile crescent" and the "cradle of civilization," so it's no wonder the area would be suited for a large population. Also, if we accept the veracity of the flood event, then the Babel scenario is a logical development. It would also be reasonable to assume that if humanity had been unified up till now, they would have a common language.

---

[314] Genesis 10 actually follows chapter 11 chronologically. It was common for Hebrew writers to summarize an event, then retell it with details. Chapter 10 identifies the clans that left Babel.

The significance of the next actions is easy to miss when the meager passage is interpreted from a contemporary perspective…

> And they said to one another, "Come, let us make bricks, and burn them thoroughly." And they had brick for stone, and bitumen for mortar. Then they said, "Come, let us build ourselves a city and a tower with its top in the heavens, and let us make a name for ourselves, lest we be dispersed over the face of the whole earth." (Genesis 11:3-4).

Why was God so opposed to them building a city and forming a strong social union? When encountering unusual passages, it's always prudent to trust in God's holy character and assume that we don't understand completely. Fortunately, though, from the context and other passages, we have some of the background for the situation. God's intent for mankind from the very beginning was for humans to "be fruitful and increase in number… fill the earth and subdue it."[315] But the intent of the people here was clearly to snub this command and become one great, unified metropolis.

And remember, great-great-grandpa Noah was still around at this time. Can you imagine him sitting on the sidelines as all of these events are transpiring? I don't doubt that the "preacher of righteousness", as James described him, was actively reminding the population of God's decree.[316] Babel's plan, then, was in flagrant, intentional disregard for God. Not only did they conspire to form a great nation, but it was to be independent of the God of their father Noah. The context implies that they wanted a new identity, one that discarded the old-school God.

---

[315] Genesis 1:28.
[316] 2 Peter 2:5.

Hence the tower… This structure was obviously to be a ziggurat, a common worship structure in Mesopotamia, and a symbol of their united power. The phrase "reaches to the heavens" is a misleading translation. Obviously, the people had experienced mountains which were much higher than a tower could ever be built on a plain. Extant ziggurats are short with a flat area on top intended to be a place to meet with deities supernaturally. Yahweh God had already established his terms of meeting, and they didn't need ziggurats. Again, we see the people's desire to incorporate a god (or gods) of their choosing who would concur with their new national identity. God's assessment was justly based on the rebellion he observed…

> And the LORD said, "Behold, they are one people, and they have all one language, and this is only the beginning of what they will do. And nothing that they propose to do will now be impossible for them. Come, let us go down and there confuse their language, so that they may not understand one another's speech." (Genesis 11:6-7).

An omniscient God could foresee the consequences of the current trajectory. Interestingly, he used language to correct the course of his children, a gift given only to the humans of Planet Earth. This was a merciful move considering the rebellious and sinful bent of the recipients. The inability to communicate and likely the fallout from the event halted any future plans. Secular anthropologists naturally balk at this explanation, given their intricate theories of language evolution. Nevertheless, when forced to choose between contrived history and facts provided by consistently accurate Scripture, the choice is easy.[317] The results were dramatic and historical…

---

[317] Once again, an individual's acceptance of a catastrophic event (language divisions) is dependent on the degree of acceptance of the supernatural.

> So the LORD dispersed them from there over the face of all the earth, and they left off building the city. Therefore its name was called Babel, because there the LORD confused the language of all the earth. And from there the LORD dispersed them over the face of all the earth (Genesis 11:8-9).

Was the tower actually built? The passage is vague as to whether the plan to construct the tower was completed or even implemented. It seems that the people of Babel stopped building the city at some point. The tower could have been at any stage of completion. Some have doubted that the number of people involved would have been sufficient to accomplish such a feat.[318] However, this depends on assumptions about a structure of unknown size and degree of completion.

An Unseen Babel

So far, we've seen a unified mass of humanity rebel against God, have their languages divided, and be scattered as the primordial nations of the earth. The account is a lesson in itself, and it provides some important elements in the Christian worldview. Obviously, supernatural intervention was involved, yet we may be looking at only the surface activities. Michael Heiser clarifies some outcomes of Babel that are often overlooked…

---

[318] "If the division of people occurred only 100 years after the Flood, there would not be many people in the world. However, the data behind the growth rates…indicate that under some scenarios It's possible to obtain a population size greater than 1,000 individuals in that much time. This occurred at all settings of minimum CBA [child-bearing age] with a minimum child spacing of 1 year, or with small minimum CBA and a minimum child spacing of 2 or 3 years. ialso possible to arrive at over 10,000 individuals with a minimum child spacing of 1 year and a minimum CBA ≤ 17, and up to 40,000 individuals with a minimum CBA of 14, although these are not likely scenarios." Robert Carter and Chris Hardy, "Modelling biblical human population growth," *Journal of Creation,* 29(1):72–79, April 2015.

> Deuteronomy 32 describes how Yahweh's dispersal of the nations at Babel resulted in his disinheriting those nations as his people. This is the Old Testament equivalent of Romans 1… wherein God "gave [humankind] over" to their persistent rebellion. The statement in Deuteronomy 32:9 that "the LORD's portion is his people, Jacob his allotted heritage" tips us off that a contrast in affection and ownership is intended.[319]

I agree with Heiser that without the proper view of Babel, Deuteronomy 32 would be enigmatic…

> Remember the days of old; consider the generations long past. Ask your father and he will tell you, your elders, and they will explain to you. When the Most High gave the nations their inheritance, when he divided all mankind, he set up boundaries for the peoples according to the number of the sons of Israel. For the LORD's portion is his people, Jacob his allotted inheritance.

Again, note the phrase "sons of Israel." Since the discovery of the Dead Sea Scrolls and other earlier translations, the consensus is that the term was originally *sons of God.* Even from the context, this would make sense, since Israel wouldn't exist for centuries. Heiser goes on to say…

> As odd as it sounds, the rest of the nations were placed under the authority of members of Yahweh's divine council. Whereas in Deuteronomy 32:8–9 God apportioned or handed out the nations to the sons of God, here we are told God "allotted" the gods to those

[319] Michael S. Heiser, "What Really Happened at the Tower of Babel?" www.blog.logos.com, November 15, 2017. Heiser gives an excellent detailed explanation in *The Unseen Realm.*

> nations. God decreed, in the wake of Babel, that the other nations he had forsaken would have other gods besides himself to worship. The other nations were assigned these lesser gods as a judgment from the Most High, Yahweh.

This may give us more information about the divine population of Heaven discussed earlier. Certain divine entities were placed in charge (to some extent) of particular people groups. Some see this as God "disowning" the rebellious humans. Each group would now have a new, but lesser "god." A rather controversial, yet otherwise coherent idea is that these assigned entities were the basis for the various gods of mythology. This belief may have support in Daniel's reference to angelic "princes" of nations.

God, however, isn't giving up on his plan to have *his own people*, despite Satan's plans to the contrary. The saga begins in the very next chapter, Genesis 12. From the rebellious throng, God selects Abram to be the progenitor of a new nation—one that would be set apart (holy) from the others. Moreover, it would be the line from which his Son would eventually be born.

The Narrative Sequence Fallacy

The historicity of the Genesis 1-11 accounts are also important in understanding a serious error in reasoning I call a *narrative sequence fallacy*. This is the idea that the dates of the actual *recordings* of an event or idea define the original sources.

For example, say a group of people witnessed the assassination of JFK—different people, same event. That night, one of the observers was interviewed for a news article, and the account was published the next day. Then suppose fifty years later, a reporter doing a "historical" article interviews another of the witnesses and creates a fresh documentary. Someone now compares the two documents and exclaims "Plagiarism! The piece published later is obviously copied from the original!"

This mistake is very common among skeptics and Bible critics who aren't aware of history or who choose to ignore it. We know that there are many ancient stories (older than Genesis) which include the first two people on earth who sin and are punished by their deity. Some of them include a garden or a snake, and they clearly mirror the biblical account. The same is true of the flood account. This may be disturbing to Christians who have been fed an exclusive diet of secularized beliefs about history.

Scripture paints a clear picture that our heritage can be traced back through the human population centered on the Plain of Shinar (Mesopotamia) after the flood, and beyond. Naturally, we would expect this monolithic culture to teach and share history to new generations. They would have a common narrative for many events such as Eden, the fall, a serpent, angels, sin, a flood, and great-great-great grandparents who built a big boat.

As the groups dispersed, these narratives would go with them. Over time it wouldn't be surprising to find the accounts deteriorated or purposely embellished and altered to fit the new culture. Thus, when Moses finally wrote his version of the histories, it would carry the same foundational truths.

Now obviously, as a conservative evangelical, I believe that Scripture is inspired. Is it so incredible that God would preserve (supernaturally if necessary) a true rendering of events through to the mind of Moses? We would expect this accurate account to have many similarities to the lore of other groups, as well as some differences. The similarities become very significant then, since they provide corroborating evidence of the same event. Some of the earliest Mesopotamian/Canaanite literature quite often describes events similar to the Old Testament. *Gilgamesh,* in particular, contains many biblical parallels. Rather than stories borrowed from a preexisting source, as skeptics accuse, the biblical narratives are inspired accounts of the same incidents. These parallel accounts actually provide external confirmation of the credibility of Scripture…

> Old Testament appropriation of Canaanite culture is a case of subversion, not syncretism—overthrowing cultural narratives as opposed to blending with them.[320]

[320] Brian Godawa, "Old Testament Storytelling Apologetics," www.equip.org, Article ID: JAF3343, 8 Mar 2015.

22
Holy Ground

"'Do not come near; take your sandals off your feet, for the place on which you are standing is holy ground.' And he said, 'I am the God of your father, the God of Abraham, the God of Isaac, and the God of Jacob.' And Moses hid his face, for he was afraid to look at God." Exodus 3:5-6

My mother taught me well—"No running, no eating, no chewing gum—this is God's house." I still feel the "specialness" when I walk through the door of that country church. Recently, I heard an old pastor bemoaning the trend in many evangelical churches of replacing ornate worship areas and stained glass with steel buildings and spotlights. I have mixed feelings…

On one hand, I realize that the early Church certainly didn't utilize grand cathedrals. Like many of the "house churches" in the world today, they met wherever they could, be it a home or a cave. Our doctrine is clear—the Church is a body, not a building. Yet like the old pastor, another part of me misses the "sacredness" of a place of worship.

There's a fine line separating what one side would consider idolatry and what the other sees as irreverence. I believe a healthy view wouldn't compromise either of these opinions by attempting to meet on middle ground. Instead, it would incorporate both of these truths simultaneously. No, in a biblical worldview a place or building cannot in and of itself be revered as something sacred. But, yes, a location may be prescribed specialness by association with the holy.

The terms *holy* and *sacred* may be confusing.[321] In the most accurate sense, they mean "totally different than" or "set apart from" something else. I have a toothbrush that I use for cleaning the sawdust out of my scroll saw. In all ways, it's the same as my tooth-brushing toothbrush. I could brush my teeth with the saw brush (my sawdust is relatively clean), but I don't. The saw brush has been designated by me to be set apart (holy) for a particular role.

Intrinsic holiness is found only in God. He alone is completely and utterly "different than" all that is not God. However, he is free to designate any other person or thing as holy (set apart, made special) for a particular purpose or reason. The person or object isn't any different intrinsically, but it's now "holy" with respect to its assigned role. Thus, there is no magical force within a holy thing. Any power associated with it comes strictly from the one who ascribed to it its special status.

In Exodus 3, God warns Moses "Do not come any closer…Take off your sandals, for the place where you are standing is holy ground." People had crossed this site countless times in the past, but at this moment God was manifesting himself there. Even in the seriousness of the event, the point was not that the ground was supernaturally changed, but that God's holiness had altered its status. Defaming the place ascribed by God would have been tantamount to defaming God.[322]

Thus, for Christians, the holiness of a place exists only as a direct connection to the holiness and will of a (The) divine

---

[321] Greek *hagios*, Hebrew *qadosh*; set apart, otherly, different.

[322] So, my mother's wise admonition in its full theological form would be: *God expects the place at which we worship him to be set apart for that purpose. It's simply a building, but you will act differently here than elsewhere because it now represents God's holiness.*

being. Non-Christian beliefs commonly embrace types of object-holiness. In most cases, their concept of holiness could derive from another source. Some Native American holy sites are thought to be empowered by a spirit or ancestor who is endowing a particular place with sacredness. But many belief systems attribute holiness to more of an impersonal force than an entity. Perhaps the specialness comes from an unknown physical source. This would fit the idea of paranormal.[323] Let's examine some of the sacred places scattered around the world.

Ley Lines

Since Pythagoras discovered how to handle angles and such, earth measuring (*geodesy*) has fascinated us. In the 1800s someone noticed that on maps, medieval churches, cemeteries, and other sacred sites seemed to make interesting lines and triangles. A few years later, the Germans picked up on the quirk and added even more sites. Connections among these locations were called *heilige linien*, (*holy lines*). supposedly, Nazis looked into this oddity as a justification of some of their land claims.

In 1925, British businessman Alfred Watkins wrote *The Old Straight Track* which put forth the idea that these *ley lines* may have been utilized in some way by prehistoric humans and early Romans. Ley line markers were typically an odd mix of stone circles, mounds, long-barrows, standing stones, cairns, ponds, springs, wells, hill notches, castles, churches, and forts. By the mid-1900s the concept flourished, mostly in Britain as a type of pseudoarcheology.

The first paranormal aspect conjured by this idea is how early humans knew the geometry of sites on such a large scale. Did they possess a complex science that is unknown today?

---

[323] Interestingly, it's also the same classifications we saw in magic and prophecy.

Were they able to detect strange force lines on the earth? Was there supernatural assistance available?

It does seem to be a mystery. A popular example is the near 3-4-5 right-triangle formed by connecting Stonehenge, Bryn Celli Ddu, and Arbor Low. Others show interesting geometries at the Great Pyramids, Machu Picchu, and similar sacred or historic sites. These are often hundreds, even thousands, of miles apart and separated by mountain ranges and bodies of water.

Explanations tend to be somewhat untestable (hence, pseudoarchaeology). A popular interpretation is that ETs had to be involved since either GPS or a high birds-eye-view seems to be required. We might ask, however, why aliens would want us to build geometrically. Alternatively, if there are some sort of force lines involved, they don't seem to correspond to any energy known to science.

A explanation might be a type of pareidolia in which humans tend to extract sensible images out of random patterns. It's possible (not probable) for three random points to coincidentally form a straight line or a right triangle. But the more points I add to the mix, the more possible lines or shapes are perceived. If I randomly placed five beans on a table, there is a small possibility that they may form a rough shape of something. If I spill 1000 beans on a table though, I could easily connect *selected* beans to form the face of William Shatner.

Stonehenge

One of the most mysterious and popularized archeological sites in the world, Stonehenge has close to a million visitors each year. The present remains are the result of over 5000 years of sporadic work. Since this would involve multiple generations (and perhaps even multiple cultures) the objectives of the builders are difficult to discern.

The first artifact at the site was simply a circular bank-and-ditch structure (*henge*) approximately 360 feet in diameter dated to about 3000 BC. The chalky material dug from the ditch was thrown to the outer side to form the bank. A large unexcavated section of the circle, perhaps a doorway, was located on the northeast side with a smaller one on the southern. Just to the inside of the ditch, another circle of 56 holes was dug. Each of these *Aubrey holes* was a little over 3 feet in diameter. It's speculated that these holes held wooden timbers or the tall pieces of *bluestone* located around the site.

About 500 years after the initial henge was excavated, what appear to be two concentric circles of holes were dug in the center of the site. The holes held up to 80 standing stones, 43 of which remain on-site. These bluestones were transported from over 100 miles away.

Over the next 200 years (2600 BC to 2400 BC), the familiar circle that most people think of as Stonehenge was constructed. Thirty enormous *sarsen* stones were transported to the site, possibly from a quarry about 25 miles to the north at the Marlborough Downs. These stones were sculpted with very precise mortise and tenon joints then erected as a circle of *standing stones,* 108 feet in diameter with a ring of 30 *lintel stones* placed on top. The standing stones average around 13 feet high by 7 feet wide and weighed around 25 tons. The inward-facing surfaces of the stones seem to be smoother than the outer surfaces.

Within this circle, five sarsen *trilithons* were arranged in a horseshoe shape about 45 feet across with its open end facing northeast. Each trilithon consisted of two standing stones with a straight lintel on top. The smallest of the trilithons was about 20 feet tall, and the *Great Trilithon* at the southwest end is 24 feet tall.

As with other ancient circles and henges, there has been debate over the precise placement of the stones. Fortunately, the original structure of Stonehenge is a bit more discernible than other sites for three reasons. First, the chalky gravel of the area retains signs of excavation better than some substrates. We can tell more easily where stones might have stood or where significant digging has occurred.

Also, the henge and avenue formations are most likely in the original positions since it would be difficult to move a ditch without some trace. This helps to estimate the centers and borders of the stone components. Finally, the interlocking mortise and tenon circle for the lintels gives an accurate measure for the entire original structure and possibly the missing stones. Many scientists believe that current models of Stonehenge are reliable and may be used to determine possible uses.

I've given the preceding details to emphasize the incredible amount of manpower and skill required for such a structure. More notable though, is that a major undertaking would seem to require a major motivation—one that would energize multiple generations.

The purpose of Stonehenge has been argued for centuries. Researchers agree that the position of its stones coincides with several astronomical alignments. One obvious feature is that the trilithon horseshoe and the Avenue are aligned with the summer solstice sunrise. Some researchers maintain that the entire site, along with other henges and stone circles, were primarily used as ritual or burial sites. Others suggest that Stonehenge may have been utilized in multiple ways by various civilizations over its long existence.

The complex nature of the site, the implications for its use, and our lack of knowledge fuel paranormal claims about Stonehenge. Foremost, the site is associated with the Druids.[324]

[324] Most of Stonehenge was completed before the first known records of druidry. John Aubrey in the 17th century connected the two. I think it likely that the civilization

What little we know of this pagan Celtic group comes from the descriptions by Julius Caesar after his conquest of their land and medieval Irish writings. Druidry seemed to be the quintessential pagan religion, embracing sorcery, astrology, and human sacrifice. Modern Druids are a loose mix of nature religion and New Age ideology.

The mysterious nature of druidry kindles all manner of paranormal ties to Stonehenge and other circles. Having studied some of these sites in-depth, I believe they had astronomical significance. There are too many celestial indicators in Stonehenge for it to be coincidental. These indicators suggest a highly educated group of stargazers. Whether scientific, religious, or paranormal, we may never know.

Nature Religions

From antiquity, nature religions have been a cross-cultural phenomenon. Beliefs vary widely, but a common tenet is that the natural world is the embodiment of divinity, ancestral spirits, or a type of impersonal force. Divinity could be expressed as mono- or polytheism. This force/spirit may exist *in*, *among*, or *as* objects (animals, plants, rock, sky, etc.) depending on local belief. Many indigenous peoples have elements of nature religion in their traditions. Similar practices are also common among modern groups with New Age connections.

Typically, "location" is important in these beliefs, since the particular environment is made up of the indwelled objects. A region, mountain, river, or other geographic feature may be singled out as the holy place. Burials may also be performed at the site, making it even more sacred. Once again, the erroneous idea persists that a location isn't simply revered, but is endued

---

related to the Druids were responsible for Stonehenge, lacking any other prospects. We do know that since the resurgence of druidry, the stone circles have been important holy places.

with a vague, mystical quality of "holiness." In our age of hyper-tolerance, adherents to nature religions are often held up as examples to be emulated…

> Whatever is native or indigenous is indiscriminately praised, and woe to those who have the temerity to point out any flaws in native cultures or religions.[325]

Practitioners usually have strong ties to the land and tend to care for the natural environment. This is certainly a good thing for our ailing ecosystems. But in the worldwide "green" economy, any set of ideologies that include an eco-friendly component are often accepted wholesale with little acknowledgment of other weaknesses. Sadly, evangelizing these groups is often condemned as an encroachment of evil Western traditions. Our task as Christians is to lovingly build opportunity bridges that maintain rich heritage and traditions, yet clearly hold fast to Scriptural truths.

Bermuda Triangle

While sailing near Bermuda, Christopher Columbus recorded a fiery ball crashing into the ocean accompanied by erratic compass readings. Over four centuries later, five military planes go missing, subsequent articles and books are produced, and a paranormal legend is born. The Bermuda Triangle is now probably the most talked-about paranormal region on the globe. (Area 51 may be gaining ground.) Seventy-five planes and hundreds of boats have been lost in the half-million-square-mile area.

---

[325] Dave Hunt, *Occult Invasion: The Subtle Seduction of the World and Church*, (Eugene, OR: Harvest House, 1998), p. 133.

Explanations abound for the disasters and disappearances, rivaling UFOs in frequency and diversity. Natural and supernatural explanations include weather, aliens, unusual waves, energy fields, demonic activity, curses, mud volcanoes, sea monsters, and conspiracy theories. So far, none of these is in the lead as a plausible cause.

The Bermuda Triangle mystery is an excellent case study in the development of a paranormal myth, perhaps providing a prototype for analyzing some other alleged paranormal phenomena. Weather may be involved, but another, more subtle "perfect storm" is likely the culprit. I believe a convergence of several factors has led to the notoriety of the area…

1. A notable historic connection. Columbus's encounter alone didn't draw much attention. It was in hindsight that historians noted the occurrence.

2. A highly publicized contemporary incident. In 1945, five bombers (Flight 19) lost contact with their base while over the area. Ultimately, all were lost along with a rescue plane.

3. Other historic events connected. With the mysterious disappearance of Flight 19, several prior disasters were called into remembrance: In 1918 the cargo ship *USS Cyclops*, 1919 the *Carroll A. Deering.*

4. Subsequent supporting events. Now that an embryonic legend is developing, new events are correlated with the former. 1948 Douglas DC 3, 1949 *Star Tiger* and *Star Ariel*, 1955 *Connemara IV.*

5. A large sample area with many data points. The Bermuda Triangle is a very large area in the very heart of major shipping lanes. Statistically, we would expect a larger number of incidents (although still proportional to other areas).

6. Increasing publicity, decreasing fact-checking. Mystery draws attention, and over the past 70 years, a plethora of pop media productions have accumulated. As we are now witnessing at light-speed on the Internet, information overload often leads to zero accountability.

I suspect that several different paranormal claims derive from similar convergence of evidence. The facts related to the Bermuda Triangle phenomena aren't that difficult to locate, even from original sources. Interestingly, insurance giant Lloyds of London hasn't found actuarial risks significantly higher in the area, which says a lot. When we examine the individual "mystery" cases involving missing and crashed craft, there are few if any with no natural explanations. Investigations of Flight 19 and others reveal clear instances of pilot or mechanical error. This is typically overlooked by the pop media.

Nevertheless, the Bermuda Triangle has a life of its own now. I suppose most cruisers and fliers will still hold their breath while passing through. On the up side, it never hurts to exercise more safety than needed.

Vortexes

Late one night while driving alone on a secluded highway, I sensed something was wrong. I turned down the radio to hear the low buzz that seemed to originate above the car. A dizzy, disoriented feeling moved over me. Could it be a vortex? No, as it turned out I should have known better than to eat the last deep-fried chili-cheese dog on the convenience store rack.

This description of a vortex is about as good as any other. I couldn't find a definitive study on vortexes (or *vortices* for the academics). Supposedly, these are unusual energy fields found at various locations or, more mystically, a site where "the veil between worlds is thinner." The concept seems universal in human cultures—a mysterious place where assorted paranormal events are more likely to occur or are more frequently reported. Some New Agers and psychics claim the ability to sense vortexes. Many are said to be regenerative, leaving one refreshed, energized, and in tune with the cosmos. Thousands visit the alleged Sedona Vortex annually for that very purpose. A Sedona website touts "It is a cathedral without walls. It is Stonehenge not yet assembled."

What is the energy behind a vortex? Depending on whom is asked, it could be natural energy such as electromagnetism and gravity, or paranormal/supernatural. Alleged paranormal sources are ley lines, chi, or spirit energy. Those leaning more toward the physical believe a vortex could be an interdimensional portal connecting two worlds. Almost every paranormal phenomenon has been associated with vortexes in one way or another—Stonehenge, the Bermuda Triangle, sacred sites, UFO hotspots, ghosts, and even Bigfoot appearances.

One popular vortex is a tourist stop in Gold Hill, Oregon. Historically, the area was avoided by Native Americans as a cursed site. Early reports also tell how horses would refuse to go near the location. In the early 1900s, an assay office located there shifted on its foundation and landed in a slanted position. Shortly after, the building changed hands and the new owner claims to have witnessed paranormal events, mostly related to gravity and disorientation (probably not unusual in a tilted house). In the 50s, the building became a destination for paranormal-pursuing tourists. Visitors are intrigued by strange phenomena such as size distortions, dizziness, camera aberrations, and the occasional glowing orb. Like most other vortexes, this one is invisible to senses and instrumentation, thus believers are at the mercy of psychics and indirect evidence.

I was able to examine a local vortex recently. At the base of a small mountain in southern Oklahoma, lies a stretch of road where out-of-gear vehicles supposedly roll uphill. I tested the phenomena several times with my truck. It did roll right up to the base of the mountain, seemingly against gravity. The scientist side of me said it was an optical illusion caused by the mountain in the background. Unfortunately, the wimpy side of me said it was getting dark and the place was creepy, so I didn't get to do any further analysis.

One of my life rules is to never visit a paranormal site that charges admission. I haven't been able to visit any other free sites to do my own research, but I'm almost certain there's a vortex in the back of my sock drawer. I've experienced several years of losing one of each pair. A few weeks ago, I remedied the situation by throwing away all the unmatched socks and buying five pairs of plain, black, uni-foot socks. I'm down to seven socks now, but at least they all match.

Crop Circles

Circular shapes in fields have been reported for centuries, the usual culprits believed to be whirlwinds or dust devils. It's also common to find circles of plants that appear different from the surrounding ones due to soil variations. Mushroom fungi can produce interesting formations called *fairy rings*. As the underground fungal hyphae spread, the sporulating mushrooms pop up on the perimeter making neat, but mysterious circles. Legend says that if you step within a mushroom circle, you'll be kidnapped that night by either fairies or the devil. I tested it with my cat, and she's still here.

Inspired by such oddities, two Brits began covertly creating elaborate crop circles in 1978. They claim to have made hundreds around the English landscape as the legend grew. Soon, copycat circles were appearing worldwide. Some of the more detailed and beautiful crop circles are made with GPS and transits. Paranormalists jumped on the new phenomenon pointing to demons, ghosts, UFOs, and strange energy fields.

Interestingly, some circles do have a paranormal connection of sorts. A group of artists calling themselves Circlemakers is responsible for many of the most intricate crop circles. There have been a significant number of incidents in which observers experience physical ailments at the sites. Gary Bates writing about Circlemakers notes that…

> …before they entered the lucrative commercial market, they called themselves Team Satan. This is a very bold name and might be suggestive of their intent.[326]

Circlemakers admits…

> Our crop formations are intended to function as temporary sacred sites… We are certain that our artworks are subject to the attention of paranormal forces and act to catalyze other paranormal events.[327]

In other words, the claim is that circles cause or attract the paranormal. As a simple experiment, I mowed a rather intricate pattern in my backyard (next to the mushroom circle). So far—nothing except perplexed neighbors.

---

[326] Gary Bates, *Alien Intrusion: UFOs and the Evolution Connection*, (Powder Springs, GA: Creation Book Publishers, 2018), 218.
[327] John Lundberg and Rod Dickinson, "Press release: Crop Formations: We Are Artists, This Is Our Work," www.circlemakers.org, 1997.

Feng Shui

Pronounced "fung shway," this is another Eastern mystical practice that has nestled itself comfortably into the American tradition. Some of the largest corporations and wealthiest families have gone to great lengths in designing work and home spaces that channel spiritual "energy" freely and productively.

Originating in Chinese Taoism, feng shui is the practice of placing architectural features and furnishings in a way that facilitates the flow of *chi* (or *ki*) energy. In Taoist cosmology, chi is the force that binds all else together. The "wrong" arrangement of physical objects (in homes and nature) can hinder the smoothness of chi currents.

Some random feng shui guidelines are…

1. Avoid houses situated near power lines, cemeteries, and junkyards.
2. Don't align the front door with the back or chi will flow right through.
3. Close toilet lid before flushing to remain wealthy.
4. Put a mirror on the inside of the bathroom door to keep in negative chi.
5. Never have the kitchen door facing a bedroom.
6. Place beds diagonal to doors and windows.
7. Oil squeaky hinges and knobs which divert chi.

Feng shui is a good model for studying the appeal of Asian philosophy to westerners. Asians' rich traditions have given the world many useful ideas and practices. The biblical worldview and empirical science, however, are very much at odds with Eastern metaphysics. Westerners (many Christians included) who would otherwise reject folklore and myth, seem to ignore illogical and unsubstantiated tenets carrying the far-east stamp of approval.

At best, feng shui seems to be a waste of effort since no scientific studies have correlated its practice with its alleged outcomes. At worst, it's a misleading substitute for biblical principles governing our lives. Instead of placing faith in the providence and goodness of God, mere dabbling in practices such as feng shui open the door to pagan concepts, a tainted witness, or possibly worse.

Spiritual Geography

> "I pushed the enormous plank door open just enough to squeeze inside. Immediately, the massive staircase seemed to open its mouth wide and breathe me up the dusty steps. A certain evil lit softly on my back prodding me onward. The ectoplasmic force was as palpable as any solid object I had ever touched…"

Yes, this is possibly some of the worst Gothic fiction ever written, but for a moment, look past the slush. A common theme, especially in ghost stories is the haunting of a *place*. Do ghosts prefer certain locations? Or in a broader sense, do certain sites "attract" paranormal activity? [328]

Superficially, this seems to be the same idea as a *holy* or *sanctified* place. Recall, though, that a place or object that is made holy has no "mystic force" of holiness applied to it. On the contrary, the place has been "set apart" for a particular purpose *by* a holy being (namely God). There is no intrinsic holiness in a location and no being is believed to be "attached" to it. For example, the Holy of Holies in the temple was not necessarily *linked* to God but was *designated* special by God for the purpose of meeting with humans. Its sacredness was not

---

[328] This will also apply to relics. Can certain objects become "unholy" or associated with a supernatural entity? The question is often asked regarding charms, talismans, and idols.

inherent but was representative of the God who had sanctified it.

The idea of "unholy" geography doesn't seem to fit this pattern. Proponents of *territorial spirits* generally understand this phenomenon to be fallen spirits that are somehow attached to a location or who have made it their residence. It isn't simply set apart to be *unholy* in the same way that God makes a place *holy*.

The book of Daniel speaks of at least two fallen spirits who seem to be assigned to earthly sites. Also, in Revelation 2, Jesus spoke of Pergamum as the place of "Satan's throne" and of the synagogue at Smyrna as "belonging to Satan."

Several years ago, a prominent evangelist organized a series of prayer activities to bind or drive out territorial "demons of perversion" that he claimed had settled in San Francisco. I applaud anyone who acts on the call to immerse an area in prayer. The nuances of how we stand against Satan should never become a dividing point in the Church. However, I do believe it's to our advantage to "fine-tune" our thinking and praying to better coincide with the realities revealed in Scripture.

It's worth noting that while numerous demons were cast out of individuals, Jesus and the apostles never confronted territorial spirits. Once again, the Ephesians Hypothesis is applicable. After listing some of the demonic categories (it's possible that some of these are territorial spirits) Paul lays out the battle plan—*put on your armor* and *stand.* Do these components of armor seem a bit familiar? Righteousness, honesty, eagerness to spread the gospel, a clear doctrine of salvation, knowledge of the Scriptures… Of course, it's simply traditional spiritual discipline! The Christian's armor isn't so much a unique incantation for exorcism as it is an encouragement to maintain and practice holiness, spiritual growth, and faith. But Paul isn't finished. He adds one more crucial element…

> …praying at all times in the Spirit, with all prayer and supplication. To that end, keep alert with all perseverance (Ephesians 6:18).

Obedience and spiritual discipline prepare us, but the power for opposing the demonic is *God's power* through our prayers. Although binding and rebuking Satan may support our *attitude* against evil, it may tend to pull resources away from biblical strategies. I believe the most efficacious battle plan is to build a solid relationship with Christ and exercise biblical spiritual discipline…

> Spiritual warfare is not about naming territorial spirits, claiming ground, or binding demons… It is to live a gospel life, to preserve gospel unity and to proclaim gospel truth.[329]

It may be that fallen spirits tend to operate more flagrantly in particular places, but we'll likely discover that they aren't nearly as attached to the place as they are to the people of that place who grant them free access.

---

[329] Timothy Chester, *The Message of Prayer: Approaching the Throne of Grace*, (Downers Grove, IL: InterVarsity Press, 2003), 231.

23
An Undiscovered Country

"The fairy tale stirs and troubles him (to his life-long enrichment) with the dim sense of something beyond his reach and, far from dulling or emptying the actual world, gives it a new dimension of depth. He does not despise real woods because he has read of enchanted woods: The reading makes all real woods a little enchanted." C.S. Lewis[330]

"We see life differently when we realize that death isn't a wall but a turnstile, a small obstacle that marks a great beginning." Randy Alcorn[331]

Every "once upon a time" begs for the accompanying "in the land of..."

Psychologists and sociologists muse at our attempts to discover Oz and Wonderland. Their theories have us groping wildly for any place offering protection from a world that has become too sad or dangerous. But this quest for magical lands wasn't strapped onto our psyches as an afterthought. Our Creator not only placed us in a Story, he made us perfectly suited for its setting.

---

330 C.S. Lewis, *Of Other Worlds: Essays and Stories.*

331 Thanks to Randy Alcorn for informing my view of Heaven. Randy Alcorn, *Heaven*, (Carol Stream, IL: Tyndale, 2004), 463.

Christians are so often accused of creating a fairytale afterlife as a crutch for hobbling through a life filled with trials. To the contrary, my entire being is tailor-made to fit into another realm—my true home. A loving and good God would do nothing less! Like any lost creature, our search for that perfect place is merely an outgrowth of the longing we feel for our natural habitat. The writer of Hebrews speaks of this yearning…

> [Those who have already died in the faith] acknowledged that they were strangers and exiles on the earth. For people who speak thus make it clear that they are seeking a homeland. If they had been thinking of that land from which they had gone out, they would have had opportunity to return. But as it is, they desire a better country, that is, a heavenly one. Therefore God is not ashamed to be called their God, for he has prepared for them a city (Hebrews 11:13-16).

But thoughts of a perfect place sometimes conjure serpents… I've dealt primarily with teenagers and young adults over the past forty years. Heaven isn't a common word in their posts and tweets.[332] I'm not too old to remember those feelings. Clearly, God blesses young people with optimism and anticipation for their future. Heaven and end-times events are often equated with death. Popular views of heaven imply the termination of these hopes and dreams in people of all ages. The consensus is that when this life ends, the party's over.

Moreover, thoughts of a sublime afterlife are a bit too close to the paranormal for "enlightened" people. In our distancing from the supernatural, we've also pushed aside one of the most uplifting teachings in Scripture. Our official doctrines of

---

[332] I capitalize "Heaven" in this chapter just to emphasize that it's a real place (my future address), as real as Shawnee, Ok.

Heaven are mostly intact, but the inward reality of a future residence with our Creator has been stifled.

This is a recent trend, though. I believe we still feel the desire for the biblical heaven, but our postmodern culture paints Heaven to be as ridiculous as a red-suited, pitchforking devil or as boring as an eternal harpfest. The idea of heaven has withered into the Christian's blind hope that there must be something better, coupled with the fear that it may not meet our 21st-century expectations. We've learned so well to misquote Paul's hope: "To live is okay, but to die is to lose everything fun."

Heaven may be the most difficult supernatural concept for my earth-realm brain to understand. Heaven itself is another realm—one that encompasses super-nature. Unquestionably my future glorified body will come equipped with a glorified cerebrum than can begin to comprehend eternity, interdimensionality, and deity. For now, though, I'm limited to only shadows of reality. Even if I could see the unseen realm now, my little three-pound bundle of neurons would go bluescreen. If God told me *white* I "would translate it into terms of dingy gray"[333]

Some have suggested that the Scriptures have little to say about Heaven. That may be true proportionally, but God's Word provides plenty of light to discern *wrong* notions. Let's examine some of the supernatural splendor of our future home by addressing some Heavenly misconceptions…

1. In Heaven, we'll give up our earthly relationships. The *Toy Story* movies tell of the relationship between the boy Andy and his cowboy doll Woody. In *Toy Story 3*, Andy is grown up and headed for college. The tear-jerking message is that Andy still loves Woody as much as ever, but other things are now more important.

---

[333] A.W. Tozer, "The Holiness of God," Sermon.

"Every good and perfect gift is from above, coming down from the Father..."[334] To imagine God would wrench away good gifts in order to make us fit for Heaven is one of Satan's most subtle lies. It isn't only contrary to the *nature* of our Heavenly Father, it's the antithesis of God's revealed *intent* to love us throughout eternity.

You will still be *you* for all eternity. The idea of nameless, amorphous spirits floating around the clouds is nowhere in Scripture. *You* will remember *your* life and relationships...

> For I know that my Redeemer lives, and at the last he will stand upon the earth. And after my skin has been thus destroyed, yet in my flesh I shall see God, whom I shall see for myself, and my eyes shall behold, and not another (Job 19:25-27).

And with perfected insight, even the sorrowful events will be seen from God's just perspective. Relationships and memories will be painless when viewed in their rightful context...

> He will wipe away every tear from their eyes, and death shall be no more, neither shall there be mourning, nor crying, nor pain anymore, for the former things have passed away (Revelation 21:4).

In Heaven, our frail, earthly shadows of love will be engulfed in perfect love. *We will not love anyone less*, but we will love everyone much, much more. I don't believe God will take away any of the special relationships with individuals. That love for your spouse, your children, and your parents will never be diminished. Rather, it will be augmented with a purer and deeper love than we could have imagined here on earth.

---

[334] James 1:17

2. I'll lose all my toys in Heaven. Perhaps that sounds too childish… I'll rephrase it more maturely, although it retains the bite: *When I go to Heaven, I'll give up all the physical things I enjoy in this world.*

"You can't take it with you…" How often do we hear this? As they say, we never see a U-Haul behind a hearse. This type of thinking, while true to some degree, doesn't correspond with biblical teachings about our eternal home. No, our bank accounts and credit lines won't go with us, but we must keep in mind God's view of his creation. His first appraisal of each aspect was "It is good." He later reminds, "Every good thing" comes from him. The evil Screwtape acknowledges this when speaking to Wormwood, his demon "trainee…"

> Never forget that when we are dealing with any pleasure in its healthy and normal and satisfying form, we are, in a sense, on the Enemy's [God's] ground. I know we have won many a soul through pleasure. All the same, it is His invention, not ours.[335]

The creation is presently fallen and corrupt, yet God hasn't given up on it. Our fallen minds perceive things in a fallen way, but that doesn't erase the original intent. Food is meant to be savored; our gluttony taints it. Peace and restfulness are gifts to the laborer; we extrapolate them into laziness. Possessions are intended for pleasure in giving, receiving, and sharing; we hoard and covet. Creative pursuits are designed to reflect the image of our Creator; we use them to invent new ways to sin. *There are no pleasures that are intrinsically sinful.* Every lustful thought, every evil act, every sinful passion is only a corruption of something good that God has ordained.

Thus, why would we expect God to remove his created goodness in Heaven? The short answer—he won't. Scripture is

[335] C.S. Lewis, *The Screwtape Letters,* (New York: HarperOne, reprint edition, 2015).

replete with God's plan to redeem the universe, not destroy it. A "new Heaven and new Earth" is the divine idea…

> Behold, I will create new heavens and a new earth (Isaiah 65:17.

> We are looking forward to a new heaven and a new earth, the home of righteousness (2 Peter 3:13).

> Then I saw a new heaven and a new earth… (Revelation 21:1).

In case you missed it, God's future plans involve not just heaven, but *earth*. A new earth, in that it's redeemed, restored, and revitalized into what God had originally willed. Randy Alcorn summarizes…

> We won't go to Heaven and leave Earth behind. Rather, God will bring Heaven and Earth together into the same dimension, with no wall of separation… Christ died not merely to make the best of a bad situation. He died so that mankind, Earth, and the universe itself would be renewed to forever proclaim his glory.[336]

What exactly will Heaven be like? Well, the "exactly" part is exactly what we won't know until we get there. That certainly doesn't mean we don't know anything about our future home. By the testimony of God's Word we know that…

> What no eye has seen, nor ear heard, nor the heart of man imagined, what God has prepared for those who love him (1 Corinthians 2:9).

---

[336] Randy Alcorn, *Heaven*, 88.

My heart imagines some pretty awesome things, and heaven is even better! We can also consider God's revealed character. Can you imagine the things an unbounded, perfectly creative, eternal, omniscient, omnipotent being can come up with? Can you imagine what he will provide for us motivated by unlimited grace and love? It doesn't simply stagger imagination—it defies it!

Furthermore, Scripture tells us that we'll have real physical bodies (though immortal and unmarred by sin) residing on a real physical Earth. I enjoy harp music. I may even devote a few centuries in Heaven to learn how to play. Maybe I'll even sit on a cloud and pluck out a little Bach for a while. But only as I wish to…

What I really enjoy here on Earth now is hiking and exploring nature. I love finding a new something over each hill. So far, I've only covered a pittance of what's out there. Will God fulfill my desire to pursue this joy in my eternal home? Perhaps that's the wrong question. I'm still approaching Heaven from a limited, earthbound perspective. I have to remind myself that God has placed me in a Story—no wait—*he created me to be in a Story.*

As mentioned previously, Heaven isn't an afterthought with features and experiences geared to match my desires like an eternal *Fantasy Island* episode. Everything about my being was planned and custom-created for my eternal existence with God *in his Heaven.* My ultimate purpose (and yours) is to "glorify God, and to enjoy him forever."[337] The reason Heaven will fulfill my desires is that I have been made expressly to be there. Every longing I have deep in my soul is the absence of something that my future habitat will include in abundance. C.S. Lewis says it better from an apologetic perspective…

[337] Westminster Shorter Catechism.

> The fact that our heart yearns for something Earth can't supply is proof that Heaven must be our home.[338]

Yes, I'll spend endless days (literally!) hiking through God's universe. Like Lewis, I believe this because I have that desire hardwired into me. After exploring every niche in this galaxy, I'll go to the next. Then perhaps I'll have to find a galaxy cluster with innumerable planets suitable for hiking. Also, I won't be limited to those that are earth-like. My recreated body won't need an atmosphere or a particular temperature range! If anyone wishes to join me, look me up sometime after the initial millennium or two.

But will we *do* anything? Is Heaven simply a big vacation?

If I'm not careful, I can become very workaholic. My family says I should learn to rest, but I just don't function that way. I thrive on *doing*. Fortunately, God has blessed me with a job I enjoy. My alarm goes off after I get up most mornings.

Heaven is often portrayed as a place of rest. Scripture even states that fact. But there is also plenty of suggestions that eternity will be an active pursuit of accomplishment. I have Jonathan Edwards to thank for relieving my Heavenly work-worries…

> We are not to suppose, when the saints have finished their course and done the works appointed to them here in this world, and are got to their journey's end, to their Father's house, that they will have nothing to do… There remaineth a rest for the people of God… The most perfect rest is consistent with being continually employed.[339]

---

[338] C.S. Lewis, *Mere Christianity*, (London: Geoffrey Bles, 1952).

[339] Jonathon Edwards, "Praise, One of the Chief Employments of Heaven," Thanksgiving Sermon, Nov. 7, 1734.

Is All This Too Pie-in-the-Sky?

If you're thinking critically of me right now for being so self-centered in my excitement over mere creaturely pursuits, don't judge me yet. You're probably asking "Won't we spend eternity just worshipping God and adoring his awesomeness? Certainly! But I think we miss the mark when we try to stuff this into the box of our earthly understanding of worship.

I can only speculate about how we'll manage time and activities in Heaven. I certainly don't think there will be competition or even indecision about what we do or with whom. This is where the goodness of a loving Savior (who has deliberately been working on a specific place just for me) is all we need to know for the time being. Yes, just being in Jesus's presence for all of eternity will never grow old. But he is the one who provides a universe filled with created wonders for our enjoyment.

When I vacation with my family at a special place, there is no competition between the sights and my loved ones. In fact, the sights and activities are enhanced by sharing them with others. And our being together is simultaneously enhanced by enjoying the experiences. I share my granddaughter's thrill of finding a new flower. My wife and I share the joy of our children sharing experiences with their children. All the grandkids share the excitement of making it to the top of a mountain together... You get the point. The symbiosis of relationships, activities, locations, and discoveries are amplified a hundred-fold over a single relationship or event.

I believe that Heaven will be that way. I literally could gaze at Christ's unbounded majesty for eternity. But when someone joins me, a new way of enjoying Christ is formed and I add an aspect of excitement that wasn't there before. When Jesus and I transport to a galaxy far, far away to see a species of plant that he created, the shared joy grows exponentially. But if Jesus is with me, is he not with you? I don't have a clue how he works it out, but an omnipresent being shouldn't have a problem. We'll adore Christ and explore God's nature forever in Heaven, and we'll enjoy each other, and we'll marvel over the endless creation and do countless other things.

"Work" isn't a 4-letter word! Adam was assigned to work well before sin entered the picture. His joyous occupation was ruling over a planet. Whatever that entailed, we can be sure he was perfectly equipped and perfectly pleased. After the fall, the dignity of work didn't change, only the mode of work here in a spoiled realm.

Several passages make it quite clear that believers will reign with Christ.[340] What will that be like? I can only imagine, but I can guess it won't be sitting on a golden throne all day twiddling my thumbs. The imagery of Heavenly work carries with it animation and adventure. God is a delegator, not out of laziness, but because he loves working with his created beings (especially his adopted children).

Once again, we must trust the goodness and love of God in anticipating our future careers. Interestingly, what we do here

---

[340] 2 Timothy 2:12, Revelation 5:10, Daniel 7:27.

on earth has input as to the responsibilities we are given in Heaven. The parable of the talents and several other passages inform us that our deeds in this earthly realm factor into our responsibilities in the heavenly realm…

> Well done, good and faithful servant! You have been faithful over a little; I will set you over much. Enter into the joy of your master (Matthew 25:21).

3. Heaven should be somber. It's a serious place with a serious Boss. Some think of Jonathan Edwards as austere and constantly nose-to-the-grindstone, yet even he admits that one of his life goals was "…to obtain for myself as much happiness, in the other world as I possibly can."[341]

C.S. Lewis so aptly notes: "Joy is the serious business of heaven."[342] Scripture encourages us to actively seek joy…

> Delight yourself in the LORD, and he will give you the desires of your heart (Psalm 37:4).

He agreed heartily with Edwards' life objective. Despite popular opinion, God is okay with joy and happiness! In fact, he created us with the capacity and desire to be happy; it was meant to be our normal state. Our pursuit of true happiness isn't selfish or materialistic. Edwards could have his joy-seeking mindset because he realized that joy, delight, and happiness are ultimately bound to the glory of God himself. He explains…

---

[341] Henry Rogers, *The Works of Jonathan Edwards, A.M.*, (London: John Childs and Son, 1839).
[342] C.S. Lewis, *Letters to Malcolm: Chiefly on Prayer*, (New York: Harcourt, Brace, Jovanovich, 1963), 93.

> God is glorified not only by His glory's being seen, but by its being rejoiced in. When those that see it delight in it, God is more glorified than if they only see it.[343]

Much of John Piper's work deals with the joy of God. He sees this as a non-negotiable in understanding the Creator and his work…

> "Our God is in the heavens; he does all that he pleases" (Psalm 115:3). Therefore, he is not frustrated. He is an unshakably happy God… His happiness is foundationally the delight he has in himself. Before creation, he rejoiced in the image of his glory in the person of his Son Then the joy of God "went public" in the works of creation and redemption… All the works of God culminate in the praises of his redeemed people… The climax of his happiness is the delight he takes in the echoes of his excellence in the praises of the saints.[344]

Heaven will be a place of joy because God is joyful. Grumpy, apathetic, or grumbling people need not apply!

4. If we are too heavenly minded, we will be no earthly good. This sounds good as a song lyric, but I don't believe I've ever seen it play out in real life. Certainly, Scripture never warns against heavenly thinking…

> Set your minds on things that are above, not on things that are on earth (Colossians 3:2).

---

[343] Jonathan Edwards, *The Works of Jonathan Edwards*, vol. 13, ed. Thomas Schafer, (New Haven: Yale University Press, 1994), 495.
[344] John Piper, "An Unshakably Happy God," John Piper devotional, www.desiringgod.org, Accessed 30 Jul 2020.

Those who keep in mind their ultimate destiny have a much more accurate view of the day-to-day narrative.

This book has a shelf life. At some point in time, the supernatural will no longer exist…

In a sense, it's been wrong from page one. I hope by now the reader has realized that what we've been calling the supernatural realm is, in fact, the *real* realm. Yes, it's only semantics, but I think our terminology adds to some peoples' wrong-thinking about the nature of reality. Perhaps a better way to visualize this is that God resides in the truly Natural realm, while we now inhabit a temporary, *sub*natural world. On some future date, this "bubble" of physicality in which we now live will be "burst," and what we now perceive with only five senses will forever be swallowed up into the *real* reality…

> Let not your hearts be troubled. Believe in God; believe also in me. In my Father's house are many rooms. If it were not so, would I have told you that I go to prepare a place for you? And if I go and prepare a place for you, I will come again and will take you to myself, that where I am you may be also (John 14:1-3).

24
Deception's Dead End

"The coming of the lawless one is by the activity of Satan with all power and false signs and wonders."
2 Thessalonians 2:9

In previous chapters, we've examined some specific phenomena that many classify as paranormal. Ephesians 6 informs us that regardless of our feelings, sensory input, or philosophical reasoning, a supernatural realm exists in which there are malevolent entities engaged is a serious battle against God and his created beings. The fact that we are lulled into complacency by the "realness" of the physical world only gives the Enemy one more weakness to usurp. One of Satan's advantages is a clear view of both realms…

> We are surrounded by another dimension of reality that is not subject to the senses, but is nevertheless as real as anything material. It is capable of penetrating our dimension in given instances and of doing things contrary to the laws which govern our dimension.[345]

I feel the pain of anyone who finds the battle metaphor unfit for the Christian experience. We (the Church) have allowed this complacency to settle into our traditions and define

---

[345] Walter Martin, *Tools of the Occult*, Audio on www.calvarytruths.com, Accessed 23 May 2020.

a "gospel" that is foreign to the one for which martyrs gave their lives. Scripture speaks often of peace, beauty, and contentment, but the overall picture (for now) is one of warfare.

God may mercifully shield dutiful Christians from the fiery frontline assault much of the time, but it wouldn't be too extreme for our first prayer each morning to be praise and thankfulness that he has provided spiritual armor for that day. Like Nehemiah's wall-builders, we should go about our daily lives with optimistic joy and encouragement yet keeping a wary eye and a ready sword.

The Ultimate Paranormal Deception

I realize the danger of speculating about the future. My simple, pea-sized brain can't even comprehend why *Star Trek TOS* was canceled after three seasons and *The Bachelor* continues, much less the course of human events. We must continually guard against letting supernatural hypothesizing take precedence over the revealed Word. Owen warns of ignoring this imbalance between feelings and beliefs, and God's truth…

> Where light leaves the affections behind, it ends in formality and atheism; where affections outrun light they sink into the bog of superstition…[346]

Scripture does provide light on future events, but even here we must take care not to read into it our own bents and biases. Mature believers should prayerfully and diligently "search the scriptures daily" and "rightly handle the word of truth," never repackaging the Word to support a particular phenomenon.

---

[346] John Owen, *The Works of John Owen*, ed. William Goold, volume 1, (Carlisle, PA: Banner of Truth, 1965), 401.

Having set these lofty criteria, are we allowed to ask the question: "Will paranormal phenomena play a role in upcoming or end-times events?" My speculations are just that—*my speculations*. I'm not a prophet, but I do believe we can explore some possibilities based on logical and thoughtful observations aligned with Scriptural truths. I fully trust that the reader will wisely follow the Berean example of verifying the claims of others.[347]

The plans of the Evil One are known only to him and the Most High. He will always do whatever he deems most efficacious in achieving his ends. Tethered to God's permissive leash, Satan may change, augment, or scrap weapons as his *methodeia* requires. However, I believe it would be naïve to think the strategies that work so well in leading people astray wouldn't continue to be used in even greater ways. Among many other tactics, paranormal deceptions seem to be major weapons in Satan's warfare. Without being overly dogmatic, there are predictions in Scripture that might allude to supernatural deceit in the end-times. Paul warns that spirits and/or demons will play a role…

> Now the Spirit expressly says that in later times some will depart from the faith by devoting themselves to deceitful spirits and teachings of demons, through the insincerity of liars whose consciences are seared (1 Timothy 4:1-2).

He also speaks of "signs and wonders" performed by Antichrist…

> The coming of the lawless one is by the activity of Satan with all power and false signs and wonders… (2 Thessalonians 2:9).

---

[347] Acts 17:11

The original text could also be read "signs and lying wonders." It's unclear as to whether these are "fake" signs and wonders, or signs and wonders that lead to a false belief. Whichever is the case, people will be utterly deceived by these events, resulting in their refusal to follow the truth…

> …and with all wicked deception for those who are perishing, because they refused to love the truth and so be saved. Therefore God sends them a strong delusion, so that they may believe what is false, in order that all may be condemned who did not believe the truth but had pleasure in unrighteousness (2 Thessalonians 2:10-12).

Some accuse God of injustice here in sending this "strong delusion." Good biblical hermeneutics, however, advises us to interpret unclear verses by the clear ones. We know from many other passages that God's holy nature precludes any deceit on his part. The first and last phrases of this passage frame God's act in the middle. These people are far from innocent bystanders. The recipients of the delusion have already "refused to love the truth" and "had pleasure in unrighteousness." After snubbing God's bountiful mercies, their destination has already been chosen. In a sense, God is releasing them to their own intents, allowing Antichrist's deceptions to culminate in their ultimate end—the condemnation of the human rebels.

The important thing to note in the context of this chapter is the clear prediction that a *huge,* worldwide, fast-track, cross-cultural deception will transpire. In my opinion, this would have to be of paranormal or supernatural proportions to convince an entire unbelieving population to give allegiance to the Lawless One. Jesus gave a similar prophecy…

> For false christs and false prophets will arise and perform great signs and wonders, so as to lead astray, if possible, even the elect (Matthew 24:24).

Most biblical scholars interpret this as hyperbole meaning these signs and wonders are *so* compelling that they could *almost* fool a believer.

Some have speculated that this global "Mother of All Deceptions" isn't feasible in our global information age.[348] As we move closer to the end-times, we've already witnessed several "impossible" prophecies become old news. Prior to the 1940s, no one could imagine the nation Israel back in her homeland. Prophecies of "the whole world" viewing an event simultaneously were ridiculed mere decades ago. A cashless economic system with the power to control every transaction worldwide? This was a ludicrous prediction for most of the 20th century. We also know that because of our social and electronic connections, change can occur at light-speed. The Covid-19 pandemic demonstrated that paradigms can literally shift overnight.

We already have hints about some of Antichrist's and the false prophet's supernatural acts—fire from the sky, a resurrection, and a talking image. Let's assume, then, that the prophesied activities of these end-time spirits are of a more blatantly paranormal nature.[349] (Just a speculative mental exercise at this point.) Are there candidates in Satan's

---

[348] When science is used to test a hypothesis, we typically try to falsify the *null* hypothesis. For example, the hypothesis is put forth that Additive X will increase apple production. We would then do experiments to *falsify* the null hypothesis: "Additive X will *not* increase apple production."

A similar logic can be used to examine possible end-times events. When skeptics claim that there is no scheme that will deceive the world, that is a null hypothesis that can be tested. One way to test it rationally is to offer scenarios in which it could be possible. If the proposal is reasonable, the null hypothesis ("World deception is not possible.") has been falsified. Note, we didn't prove the hypothesis, but we did show the null hypothesis to be unsubstantiated.

[349] As opposed to Satan's supernatural manipulation of non-supernatural constructs such as politics, academia, or culture.

contemporary pool of paranormal phenomena that could potentially initiate a global deception?

Some have speculated that the UFO/ET phenomenon could be a catalyst for rapidly altering worldviews. In 1987, President Ronald Reagan mused over that possibility: "I occasionally think how quickly our differences worldwide would vanish if we were facing an alien threat from outside this world."[350]

Imagine a scenario… One morning all TV networks switch to an unexpected press conference. The president approaches the podium accompanied by a small, gray-skinned being just like we've seen on sci-fi shows for decades. Similar announcements are simultaneously broadcast from other nations. World leaders explain that earth has been negotiating with a race of extraterrestrials for several years with the goal of "breaking the news to us gently" about our new space brothers. The world experiences what is ostensibly the greatest paradigm shift of all time.

In subsequent coverage (all regular TV programming is preempted for days), the ET spokesbeing (who has learned several earth languages) astonishes audiences via global satellite feed with the of the mysteries of the universe. He (it?) praises earthlings for showing the potential for inclusion in the galactic family. The ET encourages humans to come together as one, dropping ancient mythologies and embracing the Universal Truth—the brotherhood of all beings.

He tells of how his ancestors millennia ago engineered human life on the empty planet Earth and explains how they've been watching and nurturing us covertly for so long. But now is the "New Renaissance" of human civilization. Working through the appointed liaison (Antichrist?), humankind can now reach new heights of knowledge and creativity. Needless to say, biblical truths and the old morality in any form are

[350] UN address, 1987.

suddenly obsolete. As prophecies indicate, human civilization will wholeheartedly drink in the "Lie" and follow Antichrist.

Or could Satan trigger the strong delusion by some mystic, New Age ploy? If the rapture does occur just prior to this great deception, could a supernatural explanation be given for the mysterious disappearance of millions? Recall that one New Age leader has already predicted a miraculous "removal" of old-school Bible-thumpers so that the remaining humans can finally move to a higher spiritual plane. There are also predictions among some UFO religions that ETs will someday remove large numbers of humans (either to populate a new and better planet somewhere, or to purge the earth of "troublemakers").

Again, these are only interesting speculations, not disciplined exegesis. I believe, though, that a small helping of postulating about the future is appropriate. Whether end-times deception involves the paranormal is yet to be seen, but the fact remains that these same phenomena are also significant factors in shaping contemporary worldviews. Their influence is *now* as well as *then*.

Are we truly seeing signs of the end-times now? I believe it's prudent to realize that every tree starts as a seedling. Despite not knowing the exact dates of end-time events, Scripture encourages us to be watchful and to heed the developing signs. (This admonition in itself implies that signs are discernable.)

Here in Oklahoma, storms occasionally catch us off guard, but most of the time they are preceded by developing "clues." A nimbocumulus cloud begins billowing on the horizon. Minutes later, the western sky is completely dark. A few flashes of sheet lightning can be seen. Then a distant rumble of thunder shakes the plains. A gust front passes and the trees begin to bend. Time to head to the fraidy hole! Paul noted that the inklings of the last days were already developing in his day...

> For the mystery of lawlessness is already at work. Only he who now restrains it will do so until he is out of the way (2 Thessalonians 2:7).[351]

The scriptural metaphor for the approaching storm is the pain of childbirth. Having been only a bystander to the process, the typical birth pangs begin small and infrequent, then crescendo to intense and rapid. This is an important fact in considering end-time events. The *types* of events we see are secondary to the *intensification* of the events. Thus, if Christ's coming is near, it's possible that an "embryo" of the paranormal signs and wonders of the Tribulation may be observable today. Possibly, the "setting of the stage" will spill back into the days prior to the rapture.[352]

Will believers be tempted by the strong delusion? A pretribulation/premillennial assumption has the Church taken up to be with Christ before this deception. Some speculate individuals may become believers during the Tribulation. If this is the case, then these believers could be the referenced "elect" who were not led astray by the delusion. And of course, God would not judge his children by allowing them to be fooled by the lie.

I believe it's quite likely, however, that if the strong delusion is of a paranormal nature, we could be witnessing the embryonic stages of this today. It would be hard to doubt we are being primed for some type of major UFO deception, whether sooner or later. Even if Satan played out the ET

---

351 Whatever the "restrainer" is, it's preventing the full-scale plan of Antichrist even now. When the restrainer is absent, the plan goes hyperspeed. There are a number of guess about the identity of the restrainer. I believe it's the presence of the Holy Spirit in the Church. When the Church is raptured, Antichrist is now "unrestrained."

352 Note that the rapture has no specific signs. It occurs as "a thief in the night" with no prior warning. All of the end-times signs point to the Tribulation and Christ's second coming at the end of the seven years. Logically though, if the signs indicate an imminent second coming, the rapture is even nearer.

scenario tomorrow, I could see millions, even many Christians, dumping their biblical worldview for his paranormal upgrade.

Furthermore, some have suggested that since Satan isn't omniscient, he may not know yet the exact form of his end-times strategy. The Restrainer may have him "keeping his options open," working on a number of possible deceptions. Paranormal phenomena may be among many schemes ready for full-blown development.

The Believer's Supernatural Hope

> The war has come to us. Satan and the world rage against us without, and our flesh opposes us within. Some seem to believe that coming to Christ removes them from the battle, but the opposite is true.[353]

If that sounds too foreboding an end to our discussion, take heart! My hope and prayer is that as we've examined paranormal phenomena, the believer isn't only better equipped to discern and confront these deceptions, but has developed a strong feeling of praise and triumph—praise to the Author of our Story of adventure and redemption, and triumph in that he is sovereign over every realm, seen or unseen. He has given the specific promise...

> We would not be outwitted by Satan; for we are not ignorant of his designs (2 Corinthians 2:11).

Our "prime directive" remains clear and unwavering...

---

[353] Mark Hitchcock, *101 Answers to Questions About Satan, Demons, & Spiritual Warfare*, (Eugene, OR: Harvest House Publishers, 2014), 167.

> Put on the whole armor of God, that you may be able to stand against the schemes of the devil (Ephesians 6:11).

The strategy is simple, but by no means simplistic. Taking on this armor requires a lifestyle that cultivates the spiritual disciplines of prayer, Bible study, repentance, and purity.

Personally, I fully expect paranormal phenomena to increase in frequency, popularity, and deceptiveness. We are warned that "Evil people and impostors will go from bad to worse, deceiving and being deceived."[354] It may well be that the plot of our Story darkens before the final pages. Yet as believers, we have a sure hope…

> Who shall separate us from the love of Christ? Shall tribulation, or distress, or persecution, or famine, or nakedness, or danger, or sword?... No, in all these things we are more than conquerors through him who loved us. For I am sure that neither death nor life, nor angels nor rulers, nor things present nor things to come, nor powers, nor height nor depth, nor anything else in all creation, will be able to separate us from the love of God in Christ Jesus our Lord (Romans 8:35, 37-39).

---

[354] 2 Timothy 3:13

Chapter One?

"Biblical Christianity is not only a faith that involves essential truths; it is the story of God's purpose to redeem humanity and to bring glory to himself. This narrative is revealed to us as a comprehensive master story that is as vast as the cosmos and so detailed as to include every atom and molecule of creation." Albert Mohler[355]

"We look not to the things that are seen but to the things that are unseen. For the things that are seen are transient, but the things that are unseen are eternal." 2 Corinthians 4:18

We find ourselves at this moment in the Story, trekking a valley path connecting two mountains. If we look over our shoulder, a supernatural past is a distant peak on the horizon. We hear only faraway echoes of creation, incarnation, resurrection, and countless other miracles. Before us lies another peak shrouded in mist. At times, when our souls are very still, we seem to hear the same supernatural reverberations coming from that cloudy mountaintop.

For the moment, though, we are in between on the flatland where the rocks around us can be examined up close. The flowers can be measured, and the air can be smelled.

---

[355] Albert Mohler, (2010). "The Christian Worldview as Master Narrative: Creation, "www.albertmohler.com, 2010, Accessed 8 May 2020.

As discussed throughout this book, we valley-path believers face a double-edged danger in engaging the supernatural. Many succumb to the secularist pull to jettison the supernatural, as much as our traditions allow, in favor of a naturalistic universe. We are bombarded by this philosophy from the media, education, and too often religious leaders. Lewis saw this budding decades ago…

> They thus make it part of their method to eliminate the supernatural wherever it is even remotely possible to do so, to strain natural explanation even to the breaking point before they admit the least suggestion of a miracle.[356]

Believers are trapped in cognitive dissonance. This naturalistic philosophy works for the non-believers since it does away with any accountability beyond human preferences. But the true follower of Christ is left with an empty deism that defies both Scripture and the soul's strong sense of "something beyond."

At the other extreme, New Agers, mystics, and other pantheists urge us to abandon the physical realm completely and plunge heart-long into supernatural "realities." Ron Rhodes speaks of this end of the supernatural belief spectrum…

> Today we are living in experiential times… Rather than embracing an objective, factually-based faith, resting on divine authority, many today are seeking mystical and emotional experience… the paranormal is the experiential flavor of the day.[357]

---

[356] C.S. Lewis, "Miracles," *The Complete C.S. Lewis Signature Classics*, (New York: Harper Collins, 2002).
[357] Ron Rhodes, *The Truth*.

And therein lies the problem. Very few of the paranormal beliefs in our culture endorse the "objective, factually-based faith" detailed in Scripture. Many Christians are lured into these "empty philosophies" by Christianized vocabulary and the delusion that anything spiritual has to be of God. I don't suggest that there is a happy medium between these two extremes. Reality (God's reality as revealed in Scripture) isn't to be found on this human-centered continuum. Our gaze must be fixed higher and differently.

I like old westerns; the guy in the white hat always wins in the end. We are hard-wired for happy endings because *that is the final scene God designed for us*. Believers have the sure promise that whatever transpires in this physical realm, our Story moves to an incomparable climax in the next…

> For this light momentary affliction is preparing for us an eternal weight of glory beyond all comparison (2 Corinthians 4:17).

And even better news—that ending is only the final words of Chapter One! Eternal adventures begin in the next chapter and continue unbounded through our eternal Story with our Redeemer...

> And if I go and prepare a place for you, I will come again and will take you to myself, that where I am you may be also (John 14:3).

Carry On

I believe a solid, biblical worldview will be even more critical in the days to come. Effectively using the spiritual armor described in Ephesians 6 presupposes the warrior possesses a biblical worldview that clearly defines the enemy's methods, acknowledges the importance of the battle, and understands the resources provided.

If at this point you feel somewhat preoccupied with the supernatural or are experiencing twinges of fear, you may have missed some important points in this book. Please note these truths…

- The reality of the biblical supernatural is significant, but it's only one part of spiritual discipline and Christian maturity. Our worldview should include an accurate knowledge of the supernatural but keep a balanced view. Don't become negatively obsessed with angels, demons, or other paranormal concepts.

- Within the area of the paranormal, remember that God remains sovereign over all. For the believer, Christ's loving sacrifice has defeated sin, fear, and death. "There is no fear in love, but perfect love casts out fear."[358]

I assume that if you have read this book, you are likely a believer, concerned about scriptural truths and the biblical worldview. It's also very likely that you have a significant influence on those around you. We not only have the responsibility of sharing the gospel within our social circle but also of discipling.

Although our final destination will be glorious, Scripture tells us to expect increased deception and "birth pangs" as Christ's return draws nearer.[359] God' xs Word gives us these warnings and encouragements *because we can and should be proactive in confronting trials and deception*. Here are some

[358] 1 John 4:18
[359] "But mark this: There will be terrible times in the last days" (2 Timothy 3:1).

ways to engage your world with the truths of a biblical worldview…

1. Continue developing your own biblical worldview. In addition to the spiritual disciplines of Bible study and prayer, build an apologetic base that can provide "an answer to anyone who asks you for a reason for the hope that is in you." Place yourself (and your family) under the mentorship of those who teach Scripture as the inspired, inerrant, Word of God. Be sure they have a balanced view of doctrine and are not afraid to engage the elements of a biblical worldview. Make your beliefs accountable to your mentors and fellow learners.

2. *Use the Ephesians Hypothesis when discerning new paranormal phenomena or teachings.* Seek advice from other believers who may have experience in a particular phenomenon. Study God's Word and pray for discernment.

3. *Engage others in biblical worldview training.*[360] Provide resources to your family and others that can facilitate this. Many of the citations within this text provide good starting points for worldview enhancement.

4. *Stand against deceptive ideas.* Pray for wisdom, truth, and discernment in your family and church. One person can make a difference. "Put on the full armor!"

5. *Enjoy the reality of God's supernatural and glorify him for it!* The Most High cannot be understood to the fullest of his

---

[360] There are many good resources for this. Check out the (Christian) sources cited positively in this book. Here are some books to start with…
An Introduction to Christian Worldview (Anderson, Clark, and Naugle)
Cold Case Christianity (J. Warner Wallace)
Is God Just a Human Invention? (McDowell & Morrow)
Mere Christianity (C.S. Lewis)
The Case for Christ, The Case for Faith, The Case for a Creator, The Case for the Real Jesus (L. Strobel)

revelation without considering his supernatural attributes and creations. We are a part of his supernatural realm although presently in a limited sense. God is supernatural, yet "he is actually not far from each one of us."[361]

[361] Acts 17:27

Appendix A
Experiencing the Supernatural God

Our Eternal Creator has had the human Story in his mind forever. The Most High desires to have an eternal relationship with you forever. Nothing in your life has taken God by surprise.

The problem, however, is that we are under the curse of the Fall. We sin (fall short of God's requirements for righteousness) because we are sinners, born with a fallen nature that cannot know holy goodness. Our sin has separated us from God. We aren't able to know him personally or experience his love. Scripture tells us, "All have sinned and fall short of the glory of God" (Romans 3:23). It also says, "The wages of sin is death" (Romans 6:23). This spiritual death forces a separation from God and leaves us to continue on our way to an eternity in hell.

But because of his perfect and undeserved love, God created a way to save us by sending his son to pay the price for our sin. "God showed his own love toward us, in that while we were still sinners, Christ died for us" (Romans 5:8). This is our only hope! Jesus said "I am the way, the truth, and the life. No one comes to the Father except through Me" (John 14:6).

It isn't enough that we know these truths, we must individually place our trust in Jesus Christ as our Lord and Savior. By turning from our sins and trusting Christ's sacrifice, we can know God personally and experience His love. "For God so loved the world, that he gave his only Son, that whoever believes in him should not perish but have eternal life" (John 3:16).

You can receive Jesus by faith "If you confess with your mouth the Lord Jesus and believe in your heart that God has raised Him from the dead, you will be saved. For with the heart, one believes unto righteousness, and with the mouth confession is made to salvation" (Romans 10:9-10).

An expanded version of *The Ephesians Hypothesis*...

God and the Paranormal offers a more in-depth look at the paranormal with additional chapters such as:

RELICS AND RITUALS
The Shroud of Turin
The Ark of the Covenant
The Cross of Christ
Miscellaneous Body Parts
Exorcism

ELVIS ON THE GRASSY KNOLL
Conspiracy Theories
Gnosticism
Apocrypha and Pseudepigrapha
Bible Codes
The Illuminati
Knights Templar
The Deep State
The Real Conspiracy

SUPERMEN
The Nephilim
Giants in the Land
The Watchers
Human Hybrids
Ahead of Their Time
Other Gods

APOCALYPSE NOW?
The Bad News
The Real Apocalypse
Which View?
Coming Attractions
Prophecy Cancellers

ENCHANTED ELIXIRS AND PARANORMAL POTIONS
Alternative Medicine
The Natural Aspect
The Energy Aspect
An Alternative to the Alternative

THINGS THAT ARE JUST WEIRD

WHAT IS A WORLDVIEW?
The Adventures of Mary
Metanarrative
The Anatomy of a Worldview

SCIENCE AND THE SUPERNATURAL

PARANORMAL GLOSSARY

A novel...

Apologetic study resources for THE LIGHTS OVER MIDDLE CREEK are available at www.roserockmedia.net

Paul McLane thinks his life is finally scrutinized, analyzed, and categorized—the least that should be expected by an aspiring psychology student. But the unearthly lights in the forest behind his house make this fall's college plans a distant concern. Should he involve his grandmother? Are they in danger?

As his preoccupation with the phenomenon grows, Paul discovers his experience isn't unique. Others have witnessed the mysterious lights, and disturbingly, the death of his mother and grandfather appear to be connected to them. But the sudden appearance of Nathan Monroe sets a different trajectory for the young man's quest. Monroe's ideas about science, philosophy, and reality in general are quite intriguing. However, as Paul's worldview is evolving, God is positioning his own players for a spiritual confrontation.

Paul's obsession with knowing the truth leads him through crisscrossing roads of science, Celtic folklore, New Age mysticism, and Christian apologetics. He gives in to the consuming urge that demands answers at any cost and becomes a wounded fugitive with no place to run except to Nathan Monroe. Only God's mercy and providence, set in motion by the prayers of his faithful, can shine spiritual light over the truth about Paul's past and the hope for his future.

Made in the USA
Monee, IL
04 November 2026

67375755R00197